NONLINEAR OPTIMIZATION WITH ENGINEERING APPLICATIONS

T0156029

Springer Optimization and Its Applications

VOLUME 19

Aims and Scope
Optimization has been expanding in all directions at an astonishing rate during the last few decades. New algorithmic and theoretical techniques have been developed, the diffusion into other disciplines has proceeded at a rapid pace, and our knowledge of all aspects of the field has grown even more profound. At the same time, one of the most striking trends in optimization is the constantly increasing emphasis on the interdisciplinary nature of the field. Optimization has been a basic tool in all areas of applied mathematics, engineering, medicine, economics and other sciences.

The Springer Series in Optimization and Its Applications publishes undergraduate and graduate textbooks, monographs and state-of-the-art expository works that focus on algorithms for solving optimization problems and also study applications involving such problems. Some of the topics covered include nonlinear optimization (convex and nonconvex), network flow problems, stochastic optimization, optimal control, discrete optimization, multi-objective programming, description of software packages, approximation techniques and heuristic approaches.

NONLINEAR OPTIMIZATION WITH ENGINEERING APPLICATIONS

By

MICHAEL BARTHOLOMEW-BIGGS
University of Hertfordshire, UK

 Springer

Michael Bartholomew-Biggs
Department of Mathematics
University of Hertfordshire
Hatfield AL10 9AB
United Kingdom
m.bartholomew-biggs@herts.ac.uk

ISBN: 978-1-4419-4621-8   e-ISBN: 978-0-387-78723-7
DOI: 10.1007/978-0-387-78723-7

Mathematics Subject Classification (2000): 90-01, 90-08, 65K05

Cover illustration: by Howard Fritz: "Slipway", water-colour, 1993

Printed on acid-free paper

9 8 7 6 5 4 3 2 1

springer.com

Contents

Preface

This book, like its companion volume *Nonlinear Optimization with Financial Applications*, is an outgrowth of undergraduate and postgraduate courses given at the University of Hertfordshire and the University of Bergamo. It deals with the theory behind numerical methods for nonlinear optimization and their application to a range of problems in science and engineering. The book is intended for final year undergraduate students in mathematics (or other subjects with a high mathematical or computational content) and exercises are provided at the end of most sections. The material should also be useful for postgraduate students and other researchers and practitioners who may be concerned with the development or use of optimization algorithms. It is assumed that readers have an understanding of the algebra of matrices and vectors and of the Taylor and mean value theorems in several variables. Prior experience of using computational techniques for solving systems of linear equations is also desirable, as is familiarity with the behaviour of iterative algorithms such as Newton's method for nonlinear equations in one variable. Most of the currently popular methods for continuous nonlinear optimization are described and given (at least) an intuitive justification. Relevant convergence results are also outlined and we provide proofs of these when it seems instructive to do so. This theoretical material is complemented by numerical illustrations which give a flavour of how the methods perform in practice.

The particular themes and emphases in this book have grown out of the author's experience at the Numerical Optimization Centre (NOC). This was established in 1968 and its staff (including Laurence Dixon, Ed Hersom, Joanna Gomulka, Sean McKeown and Zohair Maany) have made important contributions in fields as diverse as quasi-Newton methods, sequential quadratic programming, nonlinear least squares, global optimization, optimal control and automatic differentiation.

The computational results quoted in this book have been obtained using a Fortran90 module derived from the NOC's OPTIMA library. This software is not described in detail but interested readers can obtain it from an ftp site. Some of the student exercises can be attempted using OPTIMA but most can also be tackled in other ways, for example via the SOLVER tool in Microsoft Excel, the MATLAB toolbox of optimization procedures or the NAG libraries in C and Fortran.

I am indebted to many people for help in the writing of this book. Besides the NOC colleagues already mentioned, I would like to thank all the mathematics staff at the University of Hertfordshire for their support. I have also received encouragement and advice from Marida Bertocchi of the University of Bergamo, Alistair Forbes of the National Physical Laboratory, Berc Rustem of Imperial College and Ming Zuo of the University of Alberta. Any mistakes or omissions that remain are entirely my responsibility. My thanks are also due to John Martindale, Ann Kostant, Elizabeth Loew and their colleagues at Springer for encouragement and help with the preparation of the book. Finally, my deepest thanks go to my wife Nancy Mattson who, for a second time, has put up with the domestic side-effects of my preoccupation with authorship.

This book seeks to capture a view of the subject that I have acquired over a working lifetime's involvement with optimization and its applications. Optimization, by definition, is concerned with making things better. It is natural, therefore, that it should apply its own principles to itself and – in my experience, at least – this can generate a lively spirit of friendly rivalry between practitioners and algorithm developers. This spirit is worth celebrating in quasi-haiku form:

Optimization
means a quest for best answers
by the best methods.

Optimism means
believing both objectives
are achievable.

I hope readers will be stimulated by the challenge of finding more and more effective solutions to practical problems that become increasingly difficult.

Michael Bartholomew-Biggs
January, 2008

List of Figures

List of Tables

Chapter 1

Introducing Optimization

1.1.   A tank design problem

In an optimization problem we seek values for certain *design* or *control variables* which minimize (or sometimes maximize) an *objective function*. A good example is the problem of finding the dimensions of a rectangular open-topped tank in order to obtain the smallest surface area which encloses a given volume, V^*. (The purpose of such a design might be to minimize heat loss through the sides.) We denote the height by x_1 and the lengths of the edges of the base by x_2 and x_3. The volume and surface area are then given by

$$V = x_1 x_2 x_3 \quad \text{and} \quad S = 2x_1 x_2 + 2x_1 x_3 + x_2 x_3.$$

Hence the design problem can be posed as

Minimize $S = 2x_1 x_2 + 2x_1 x_3 + x_2 x_3$ subject to $x_1 x_2 x_3 = V^*$.
$$\text{(1.1.1)}$$

This is a three-variable optimization problem which includes an *equality constraint*. Methods for solving problems of this kind are discussed in Chapters 16–18; but an alternative *unconstrained* formulation can be obtained by eliminating one of the unknowns. Because $x_3 = V^* x_1^{-1} x_2^{-1}$ we can also seek the optimum tank dimensions by solving

$$\text{Minimize} \quad S = 2x_1 x_2 + 2V^* x_2^{-1} + V^* x_1^{-1}. \qquad \text{(1.1.2)}$$

The solution of problems of this kind is discussed in Chapters 5–11.

The optimal tank dimensions can be found by solving either (1.1.1) or (1.1.2). However an important factor has been omitted from both of them. If any two of the x_i have negative values then the constraint

M. Bartholomew-Biggs, *Nonlinear Optimization with Engineering Applications*,
DOI: 10.1007/978-0-387-78723-7_1, © Springer Science+Business Media, LLC 2008

on volume can still be satisfied but the surface area may be negative. Because a negative value for S is necessarily less than a positive one, a solution with, say, $x_1 < 0$ and $x_2 < 0$ might seem "better" than a solution with all the x_i positive. Of course, negative dimensions have no practical meaning and so the problem formulation should explicitly exclude them. We can do this by adding *inequality constraints*, as in

$$\text{Minimize} \quad 2x_1x_2 + 2x_1x_3 + x_2x_3 \quad \text{s.t.} \quad x_1x_2x_3 = V^*, \quad x_i \geq 0, \quad i = 1, 2, 3. \tag{1.1.3}$$

or

$$\text{Minimize} \quad 2x_1x_2 + 2V^*x_2^{-1} + V^*x_1^{-1} \quad \text{s.t.} \quad x_i \geq 0, \quad i = 1, 2. \tag{1.1.4}$$

(The abbreviation "s.t." is often used instead of "subject to".) Methods for dealing with problems such as (1.1.3) and (1.1.4) are considered in Chapters 20–23.

In this chapter and the next we restrict ourselves to unconstrained problems involving only one variable. We can obtain such a problem from the tank design example by adding an extra requirement that the base must be square; that is, $x_2 = x_3$. Now the expressions for volume and surface area become

$$V = x_1x_2^2 \quad \text{and} \quad S = 4x_1x_2 + x_2^2.$$

Using the constraint on V to eliminate x_1, we get S in terms of x_2 only; that is,

$$S = 4V^*x_2^{-1} + x_2^2. \tag{1.1.5}$$

Figure 1.1 shows S as a function of x_2 when $V^* = 5$. In this case the minimum occurs when $x_2 \approx 2.2$.

Figure 1.1 illustrates the well-known fact that, at the minimum of a *differentiable* function, the slope – that is, the first derivative – is zero. Hence, for this rather simple problem, we can obtain the minimum surface area by solving

$$\frac{dS}{dx_2} = -4V^*x_2^{-2} + 2x_2 = 0$$

which gives $x_2 = (2V^*)^{1/3}$. Hence, when $V^* = 5$, the optimum square base has edges of length 2.1544.

Not all optimization problems are as easy as the minimization of (1.1.5). Some objective functions are hard to differentiate; and, even when the first derivative has been found, the equation obtained by setting it to zero may be difficult to solve. This book describes some of the computational methods used by engineers and scientists to deal with

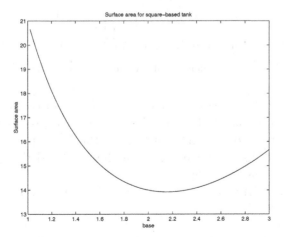

Figure 1.1. Tank surface area as a function of x_2.

optimization problems which do not have an analytical solution. Such problems occur in many situations, for example, finding a formula which gives the closest match to some experimental data, choosing the shortest route which avoids a number of obstacles, or devising a maintenance schedule which gives the least operating cost. Such case studies are used later in the book as a basis for the practical comparison of different optimization methods.

Exercises

1. What happens to the surface area (1.1.5) as $x_2 \to 0$? What is the minimum value of S if x_2 lies in the range $-1 < x_2 < 0$?

2. If $x_2 = x_3$, reformulate (1.1.2) as an unconstrained minimization problem involving x_1 only. Using the value $V^* = 5$, plot a graph of the objective function in the range $1 \le x_1 \le 3$. Hence deduce the minimum surface area. What happens to the surface area as $x_1 \to 0$?

3. Formulate the problem of finding the maximum volume that can be enclosed by a rectangular open tank with a fixed surface area and then estimate a solution when the base of the tank is square and the fixed surface area is 8. (Note that maximizing a function $F(x)$ is equivalent to minimizing $-F(x)$.)

1.2. Least squares data-fitting

Suppose that a laboratory experiment produces a record of measured temperatures, θ, (°C) against time t (minutes), as in Table 1.1. Suppose also that we believe the underlying relationship between θ and t is linear, of the form $\theta = at$, for some unknown coefficient a. The data points

Measurement i	1	2	3	4
Time t_i	1.0	2.0	3.0	4.0
Temperature θ_i	2.3	5.1	7.2	9.5

Table 1.1.   Experimental data for temperature versus time.

do not, in fact, lie on a straight line (perhaps because of experimental errors). Hence, out of all the straight lines which pass near the data points, we wish to find the one which gives the best approximation, in the sense that the discrepancies between the data and the straight line *model* are as small as possible.

Figure 1.2 shows the errors $at_i - \theta_i$ as vertical lines PP'.

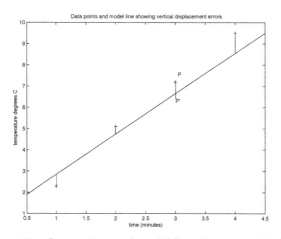

Figure 1.2.   Data points and model line showing vertical errors.

A common way to find the best approximation is to choose a to minimize the sum of squares of these vertical errors; that is, we want a to solve the problem

$$\text{Minimize} \quad F(a) = \sum_{i=1}^{4}(at_i - \theta_i)^2. \qquad (1.2.1)$$

At a minimum of F, the first derivative $F'(a)$ is zero. Hence the optimum value of a satisfies

$$\frac{dF}{da} = 2\sum_{i=1}^{4}(at_i - \theta_i)t_i = 0. \qquad (1.2.2)$$

This leads to

$$a\sum_{i=1}^{4} t_i^2 = \sum_{i=1}^{4} \theta_i t_i.$$

Substituting for t_i and θ_i from Table 1.1 we get $30a = 72.1$ and so $a \approx 2.4033$.

This simple problem is an example of the *least squares* approach to approximating a set of data points by a model function. The approach can be extended (as shown in later chapters) to models with more than one unknown coefficient.

The data-fitting problem we have just solved is extremely easy because F is a quadratic function of the variable a. This means that equation (1.2.2) is linear and yields a unique answer. We now show that some optimization problems are not so straightforward by considering another way to minimize discrepancies between the data and the model. Rather than dealing with just the vertical error at a data point, we take account of the *total displacement* given by the perpendicular distance of (t_i, θ_i) from the line $\theta = at$ as shown in Figure 1.3.

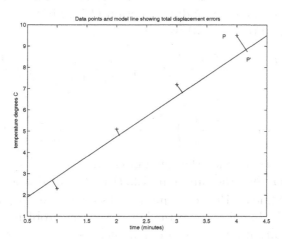

Figure 1.3. Data points and model line showing total displacement errors.

We can determine the perpendicular distance between point and line as follows. A typical point, P', on the line $\theta = at$ has coordinates (t, at) and the slope of the line joining P' to the data point P with coordinates (t_i, θ_i) is

$$m = \frac{at - \theta_i}{t - t_i}.$$

We want to find the value of t which makes $m = -a^{-1}$ because the line PP' will then be perpendicular to $\theta = at$ and P' will be the model point which is closest to the data point P. The value of t at the *footpoint* P' is found by solving

$$-a^{-1}(t - t_i) = at - \theta_i.$$

Hence P' is defined by

$$t = \tau_i = \frac{a^{-1}t_i + \theta_i}{a + a^{-1}} = \frac{t_i + a\theta_i}{a^2 + 1}. \tag{1.2.3}$$

The total displacement PP' is then $\sqrt{(t_i - \tau_i)^2 + (\theta_i - a\tau_i)^2}$ and to get the optimum straight line $\theta = at$ we must find a by solving

$$\text{Minimize } \hat{F}(a) = \sum_{i=1}^{8} \phi_i^2 \tag{1.2.4}$$

where $\phi_i = (t_i - \tau_i)$ and $\phi_{i+4} = (\theta_i - a\tau_i)$ for $i = 1, \ldots, 4.$ (1.2.5)

Of course, $\tau_1, \ldots, \tau_4$ and $\phi_1, \ldots, \phi_8$ are functions of a. If we substitute the known values of t_i and θ_i we see that (1.2.4) is a more complicated expression than the corresponding function (1.2.1) in the vertical least-squares problem. From (1.2.3) we get

$$\tau_1 = \frac{1 + 2.3a}{a^2 + 1}; \quad \tau_2 = \frac{2 + 5.1a}{a^2 + 1}; \quad \tau_3 = \frac{3 + 7.2a}{a^2 + 1}; \quad \tau_4 = \frac{4 + 9.5a}{a^2 + 1}.$$

Hence

$$\phi_1 = \left(1 - \frac{1 + 2.3a}{a^2 + 1}\right), \quad \phi_5 = \left(2.3 - \frac{a + 2.3a^2}{a^2 + 1}\right)$$

with similar expressions for the remaining ϕ_i.

It is now clear that the function (1.2.4) is not quadratic and its first derivative is not linear. Hence, forming and solving the equation $\hat{F}'(a) = 0$ is more difficult than for the vertical least-squares problem. In practice, we would normally minimize a function such as $\hat{F}(a)$ by using iterative methods of the kind described in the next chapter. More information about total least squares and the footpoint problem is given in [27].

We can, of course, estimate the minimum of $\hat{F}(a)$ by plotting a graph, as shown in Figure 1.4. In this case, the best straight line approximation in the *total* least squares sense is very similar to the approximation based on vertical least squares with slope $a \approx 2.4$.

Exercises

1. Using vertical displacements, find the straight line $y = mx$ to give a least-squares approximation to the data points $(3, 7)$, $(4, 8)$, $(6, 11)$.

2. Show that the footpoint P' could have been found by putting $\tau_i = t_f$ where $t = t_f$ solves the problem

$$\text{Minimize } (t_i - t)^2 + (\theta_i - at)^2.$$

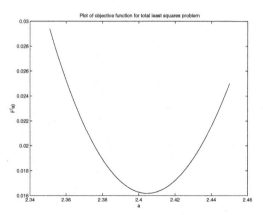

Figure 1.4. Plot of $\hat{F}(a)$ the total least squares error function.

3. Use the data in the worked example to find expressions for $\phi_2, \ldots, \phi_4$ and $\phi_6, \ldots, \phi_8$; hence complete the expression for $\hat{F}(a)$ in (1.2.4) and obtain the expression for $\hat{F}'(a)$. Plot a graph to estimate the solution of $\hat{F}'(a) = 0$.

4. Consider the data in Table 1.1 and suppose θ_4 is changed to 14.2. Calculate a model line $\theta = at$ using both vertical and total least squares. Comment on the difference between the two solutions.

1.3.   A routing problem

Suppose a robot vehicle starts at the origin and is required to proceed to a point P, as shown in Figure 1.5. It must move initially along the x-axis and then turn towards P at some point Q. The circle represents a "no-go" area which the vehicle must avoid. The point Q is to be chosen to minimize a combination of the total distance travelled and the length of the route that lies within the circle.

If the line from Q to P cuts the circle at R and S then we can define the optimum route as the one which minimizes

$$F = \text{distance OQ} + \text{distance QP} + \rho(\text{distance RS})$$

where ρ is a positive constant. This form of function penalizes the portion of the route inside the no-go region. If ρ is large we expect little or none of the optimum route to pass through the circle. On the other hand, as $\rho \to 0$, the optimum route will come closer to the straight line OP.

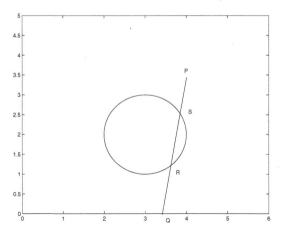

Figure 1.5.   A routing problem.

If P has coordinates $(x_p,\ y_p)$ and if x is the (unknown) distance OQ then the total length of the route is

$$d(x) = x + \sqrt{(x - x_p)^2 + y_p^2}. \qquad (1.3.1)$$

(We are assuming that P and the circle are in the positive quadrant and hence that x is positive.) We now need to determine the points of intersection (if any) of line QP and the circle. We assume that P is outside the no-go area and also that the circle does not cut the x-axis; it follows that the line segment QP will either cut the circle twice or not at all. The coordinates of any point on the line between Q and P are

$$(x + \lambda(x_p - x),\ \lambda y_p) \quad \text{where} \quad 0 \le \lambda \le 1.$$

If the no-go area has centre $(x_c,\ y_c)$ and radius r then points of intersection with QP occur when λ satisfies

$$(x + \lambda(x_p - x) - x_c)^2 + (\lambda y_p - y_c)^2 - r^2 = 0.$$

This simplifies to $\alpha\lambda^2 + \beta\lambda + \gamma = 0$, where the coefficients are given by

$$\alpha = (x - x_p)^2 + y_p^2, \quad \beta = 2[(x_p - x)(x - x_c) - y_p y_c] \qquad (1.3.2)$$

$$\text{and} \quad \gamma = (x - x_c)^2 + y_c^2 - r^2. \qquad (1.3.3)$$

We let $\delta = \beta^2 - 4\alpha\gamma$. If $\delta \le 0$ there are no points of intersection with the circle and so the distance RS is zero. On the other hand, if $\delta > 0$ the intersection points are given by

$$\lambda_1 = \frac{-\beta + \sqrt{\delta}}{2\alpha}, \quad \lambda_2 = \frac{-\beta - \sqrt{\delta}}{2\alpha}. \qquad (1.3.4)$$

The distance RS is then given by $|\lambda_1 - \lambda_2| \times$ (distance QP). Hence the optimum route is obtained by minimizing

$$F(x) = d(x) + \rho v(x) \tag{1.3.5}$$

where $d(x)$ is given by (1.3.1) and

$$v(x) = |\lambda_1 - \lambda_2| \sqrt{(x - x_p)^2 + y_p^2}. \tag{1.3.6}$$

Note that α, β and γ are functions of x because of (1.3.2) and (1.3.3). Hence (1.3.4) implies that λ_1 and λ_2 also depend on x. It is possible – but not trivial – to differentiate $F(x)$ but it will not be possible to find an analytical solution to the equation $F'(x) = 0$.

If we take the target point $(x_p,\ y_p)$ as $(5,4)$ and define the no-go region by $x_c = y_c = 2,\ r = 1$ then we can plot the function (1.3.5), as shown in Figure 1.6. It is clear that the optimum value of x is about 1.62. The solution path leaves the x-axis at a point Q such that PQ is a tangent to the circular boundary of the no-go region.

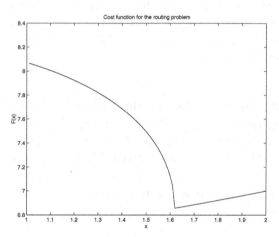

Figure 1.6. Nonsmooth cost function for the routing problem.

Figure 1.6 shows that the minimum corresponds to a "kink" in $F(x)$; that is, the slope of (1.3.5) is not zero at the optimum but instead has a discontinuity. This is due to the presence of the square root in (1.3.6). The function (1.3.5) is said to be *nonsmooth*.

Most of the optimization methods described in this book are intended for use with smooth (i.e., continuously differentiable) functions. We can formulate the routing problem in terms of a function which is smooth if we choose to minimize

$$\Phi(x) = d(x) + \rho v(x)^3 \tag{1.3.7}$$

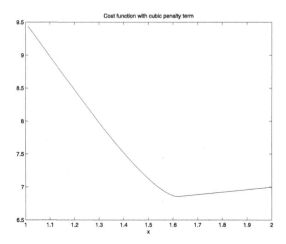

Figure 1.7.   Smooth cost function for the routing problem.

whose graph is given in Figure 1.7 (for $\rho = 1$). The minimum of (1.3.7) occurs at approximately the same place as that of the nonsmooth function (1.3.5). The use of functions such as (1.3.7) in some real-life routing problems is described in [65, 9].

Exercises

1. Calculate expressions for the first derivatives of (1.3.1) and (1.3.6).
2. Using the sample data $x_p = 5$, $y_p = 4$, $x_c = y_c = 2$, $r = 1$, plot graphs to determine the minima of (1.3.5) and (1.3.7) when $\rho = 0.5, 0.05$ and 0.005. Comment on any differences you observe.
3. Use the data $x_p = 4$, $y_p = 8$, $x_c = 4$, $y_c = 2$, $r = 2$ to plot graphs of (1.3.5) and (1.3.7) with $\rho = 1$ in the range $0 \le x \le 10$. Comment on what you observe.

Chapter 2

One-variable Optimization

2.1.   Optimality conditions

Definition Suppose that $F(x)$ is a continuous function of the scalar variable x and that, for some point $x = x^*$, there exists an $\epsilon > 0$ such that

$$F(x^*) \leq F(x) \quad \text{when } |x - x^*| \leq \epsilon. \tag{2.1.1}$$

Then $F(x)$ is said to have a *local minimum* at x^*.

If $F(x)$ is a one-variable differentiable function then we can characterize a minimum in terms of its first and second derivatives. In what follows we sometimes use the notation

$$F'(x) = \frac{dF}{dx} \quad \text{and} \quad F''(x) = \frac{d^2 F}{dx^2}.$$

Definition Suppose that $F(x)$ is a continuously differentiable function of the scalar variable x and that, when $x = x^*$,

$$\frac{dF}{dx} = 0 \quad \text{and} \quad \frac{d^2 F}{dx^2} > 0. \tag{2.1.2}$$

Then $F(x)$ is said to have a *local minimum* at x^*.

Conditions (2.1.2) are called *optimality conditions*. We have already used the optimality condition $F'(x^*) = 0$ in the examples in the previous chapter. Conditions (2.1.1) or (2.1.2) imply that $F(x^*)$ is the smallest value of F in some region near x^*. It may also be true that $F(x^*) \leq F(x)$ for all x but condition (2.1.2) does not guarantee this.

Definition If conditions (2.1.2) hold at $x = x^*$ and if $F(x^*) \leq F(x)$ for all x then x^* is said to be the *global minimum*.

M. Bartholomew-Biggs, *Nonlinear Optimization with Engineering Applications*,
DOI: 10.1007/978-0-387-78723-7_2, © Springer Science+Business Media, LLC 2008

In practice it is usually hard to establish that x^* is a global minimum and so we are chiefly concerned with methods of finding local minima.

There are stationary points of $F(x)$ which satisfy the first condition (2.1.2) but not the second. If $F'(x^*) = 0$ and $F''(x^*) < 0$ then x^* is a local maximum. But if $F'(x^*) = 0$ and $F''(x^*) = 0$ then $F(x)$ may be neither a maximum nor a minimum. For instance, the function $F(x) = x^3$ has a stationary point at $x = 0$ such that F is steadily decreasing as x approaches zero through positive values but is steadily increasing as x approaches zero through negative values.

For simple problems, the conditions (2.1.2) can be used directly to find a minimum. Consider

$$F(x) = x^3 - 3x^2. \tag{2.1.3}$$

Because $F'(x) = 3x^2 - 6x$ we have $F'(x) = 0$ when $x = 0$ and $x = 2$. Hence there are two stationary points of $F(x)$; and to find which is a minimum we must consider $F''(x) = 6x - 6$. F has a minimum at $x = 2$ because $F''(2) > 0$. However, $F''(0)$ is negative and so $F(x)$ has a maximum at $x = 0$.

We can only use this analytical approach when it is easy to form and solve the equation $F'(x) = 0$. This may not be the case for functions $F(x)$ which occur in practical problems and so we usually resort to iterative techniques.

Some iterative methods are called *direct search* techniques and are based on comparisons of function values at trial points. Others, known as *gradient methods*, use derivatives of the objective function and can be viewed as algorithms for solving the nonlinear equation $F'(x) = 0$. Gradient methods tend to converge faster than direct search methods. They also have the advantage that they permit an obvious convergence test, namely stopping the iterations when the gradient is near zero. Gradient methods are not suitable, however, when $F(x)$ is a function like (1.3.5) which has discontinuous derivatives.

Exercises

1. Show that if conditions (2.1.2) hold then $F(x^* + h) > F(x^*)$ for h sufficiently small. (*Hint:* use a Taylor series expansion.)
2. Find the stationary points of $F(x) = 4\cos x^2 - \sin x^2 - 3$.
3. Discuss the stationary points of $F(x) = x^4$, $F(x) = -x^4$ and $F(x) = x^5$.

2.2.   The bisection method

A simple (but inefficient) way of estimating the least value of $F(x)$ in a range $a \leq x \leq b$ would be to calculate the function at many points in

$[a, b]$ and then pick the one with the lowest value. The *bisection method* uses a more systematic approach to the evaluation of F in $[a, b]$.

Suppose we are seeking the minimum of $F(x)$ in the range $0 \leq x \leq 1$ and have evaluated F at five equally spaced points as shown in Figure 2.1.

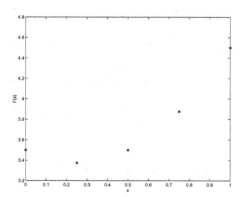

Figure 2.1. Five equi-spaced function values used by the bisection method.

If we assume that $F(x)$ is *unimodal* (i.e., that it has only one minimum in the range we are exploring) then the distribution in Figure 2.1 suggests the minimum must lie in the range $0 \leq x \leq 0.5$. (If the sample values of F had been different we might have deduced that the minimum was in one of the ranges $0.5 \leq x \leq 1$ or $0.25 \leq x \leq 0.75$.) Repeated use of this argument allows us to locate the minimum precisely, using the following formal algorithm.

Bisection Method for minimizing $F(x)$ on the range $[a, b]$

Set $x_a = a$, $x_b = b$ and $x_m = \frac{1}{2}(a + b)$.
Calculate $F_a = F(x_a)$, $F_b = F(x_b)$, $F_m = F(x_m)$
Repeat
set $x_l = \frac{1}{2}(x_a + x_m)$, $x_r = \frac{1}{2}(x_m + x_b)$
calculate $F_l = F(x_l)$ and $F_r = F(x_r)$
let $F_{\min} = \min\{F_a, F_b, F_m, F_l, F_r\}$
if $F_{\min} = F_a$ or F_l then   set $x_b = x_m$, $x_m = x_l$, $F_b = F_m$, $F_m = F_l$
else if $F_{\min} = F_m$ then   set $x_a = x_l$, $x_b = x_r$, $F_a = F_l$, $F_b = F_r$
else if $F_{\min} = F_r$ or F_b then   set $x_a = x_m$, $x_m = x_r$, $F_a = F_m$, $F_m = F_r$
until $|x_b - x_a|$ is sufficiently small

Each iteration of the bisection method compares the function values at five points in order to halve the size of a bracket containing the minimum.

We can show how the algorithm works by applying it to the problem

$$\text{Minimize } F(x) = x^3 - 3x^2 \quad \text{for } 0 \le x \le 3.$$

Initially $x_a = 0$, $x_b = 3$, $x_m = 1.5$. The first iteration adds $x_l = 0.75$, $x_r = 2.25$ and we then have

$$F_a = 0; \quad F_l = -1.266; \quad F_m = -3.375; \quad F_r = -3.797; \quad F_b = 0.$$

The least function value $F_{\min}$ occurs at $x_r = 2.25$ and so the search range for the next iteration is $[x_m, x_b] = [1.5, 3.0]$.

After re-labelling the points and computing new values x_l, x_r we get

$$x_a = 1.5; \quad x_l = 1.875; \quad x_m = 2.25; \quad x_r = 2.625; \quad x_b = 3$$

and

$$F_a = -3.375; \quad F_l = -3.955; \quad F_m = -3.797; \quad F_r = -2.584; \quad F_b = 0.$$

Now the least function value is at x_l and the new range is $[x_a, x_m] = [1.5, 2.25]$. Relabelling and adding the new x_l and x_r gives

$$x_a = 1.5; \quad x_l = 1.6875; \quad x_m = 1.875; \quad x_r = 2.0625; \quad x_b = 2.25$$

and

$$F_a = -3.375; \quad F_l = -3.737; \quad F_m = -3.955; \quad F_r = -3.988;$$
$$F_b = -3.797.$$

These values imply the minimum lies in $[x_l, x_r] = [1.875, 2.25]$. After a few more steps we have an acceptable approximation to the true solution at $x = 2$.

Proposition If $F(x)$ is unimodal and has a minimum x^* with $a \le x^* \le b$ then the number of bisection iterations needed to locate x^* in a bracket of width less than 10^{-s} is K, where K is the smallest integer which exceeds

$$\frac{\log_{10}(b-a) + s}{\log_{10}(2)}. \tag{2.2.1}$$

Proof The size of the bracket containing the solution is halved on each iteration. Hence, after k iterations the width of the bracket is $2^{-k}(b-a)$. To find the value of k which gives

$$2^{-k}(b-a) \le 10^{-s}$$

we take logs of both sides and get

$$\log_{10}(b - a) - k \log_{10}(2) \leq -s$$

and so the width of the bracket is less than 10^{-s} once k exceeds (2.2.1).

The number of iterations needed to achieve a specified accuracy depends on the size of the initial search range rather than the form of the function being minimized, as observed in the numerical examples later in this chapter.

Finding a bracket for a minimum

We now give a systematic way of finding a range $a < x < b$ which contains a minimum of $F(x)$. The method uses the slope F' to indicate whether the minimum lies to the left or right of an initial point x_0. If $F'(x_0)$ is positive then lower function values will be found for $x < x_0$, whereas $F'(x_0) < 0$ implies lower values of F occur when $x > x_0$. The algorithm simply takes larger and larger steps in a "downhill" direction until the function starts to increase, indicating that a minimum has been bracketed.

Finding a and b to bracket a local minimum of $F(x)$

Choose an initial point x_0 and a step size $\alpha(> 0)$
Set $\delta = -\alpha \times sign(F'(x_0))$
Repeat for $k = 0, 1, 2, \ldots$
    $x_{k+1} = x_k + \delta$, $\delta = 2\delta$
until $F(x_{k+1}) > F(x_k)$
if $k = 0$ then set $a = x_0$ and $b = x_1$
if $k > 0$ then set $a = x_{k-1}$ and $b = x_{k+1}$

Exercises

1. Apply the bisection method to $F(x) = e^x - 2x$ in the interval $0 \leq x \leq 1$.
2. Do two iterations of the bisection method for the function $F(x) = x^3 + x^2 - x$ in the range $0 \leq x \leq 1$. How close is F_{min} to the exact minimum of F? What happens if you apply the bisection method in the range $-2 \leq x \leq 0$?
3. Use the bracketing technique with $x_0 = 1$ and $\alpha = 0.1$ to bracket a minimum of $F(x) = e^x - 2x$.
4. Estimate how many function evaluations are used by the bisection method to reduce an initial range $a \leq x \leq b$ to a bracket with width less than 10^{-s}.

5. Discuss what will happen if the bisection method is applied in the range $a \leq x \leq b$ when $F(x)$ does not have a minimum in this range.

2.3.   The secant method

We now consider an iterative method for solving $F'(x) = 0$. This finds a local minimum of $F(x)$ provided we use it in a region where the second derivative $F''(x)$ remains positive.

Let $F(x)$ be a continuous and differentiable function and suppose $x_1 < x_2$ and also $F'(x_1) < 0$ and $F'(x_2) > 0$. Then there is a minimum of F between x_1 and x_2. A simple sketch diagram shows that this must be the case. (A similar sketch shows there must be a maximum between x_1 and x_2 if $F'(x_1) > 0$ and $F'(x_2) < 0$.) If there is a minimum between x_1 and x_2 then we can estimate its position using linear interpolation. If $F_1' = F'(x_1)$ and $F_2' = F'(x_2)$ then

$$x_3 = x_1 - \frac{F_1'}{F_2' - F_1'}(x_2 - x_1) \tag{2.3.1}$$

gives x_3 as an estimate of the point where $F'(x)$ vanishes.

The formula (2.3.1) can also be used to obtain an extrapolated estimate of a stationary point if F_1' and F_2' have the same sign. The stationary point will be a minimum under the following conditions.

either $x_1 < x_2$ and $F_1' < F_2' < 0$ or $x_1 > x_2$ and $F_1' > F_2' > 0$.

Once again a simple sketch shows why these conditions are necessary.

Consider the function $F(x) = x^2 - 3x - 1$ for which $F'(x) = 2x - 3$. If we choose $x_1 = 0$ and $x_2 = 2$ then (2.3.1) gives

$$x_3 = 0 - \frac{F'(0)}{F'(2) - F'(0)} \times 2 = 0 - \frac{-3}{4} \times 2 = 1.5.$$

In this case (2.3.1) has found the stationary point of $F(x) = x^2 - 3x - 1$. This will always happen when $F(x)$ is quadratic (see Exercise 2). When F is not quadratic, however, (2.3.1) must be used iteratively, as in the algorithm below.

Secant method for solving $F'(x) = 0$

Choose x_0, x_1 as two estimates of the minimum of $F(x)$
Repeat for $k = 0, 1, 2, \ldots$.

$$x_{k+2} = x_k - \frac{F'(x_k)}{F'(x_{k+1}) - F'(x_k)}(x_{k+1} - x_k) \tag{2.3.2}$$

until $|F'(x_{k+2})|$ is sufficiently small.

We apply this algorithm to $F(x) = x^3 - 3x^2$ for which $F'(x) = 3x^2 - 6x$. If $x_0 = 1.5$ and $x_1 = 3$ then $F'(x_0) = -2.25$ and $F'(x_1) = 9$. Iteration one gives

$$x_2 = x_0 - \frac{F'(x_0)}{F'(x_1) - F'(x_0)}(x_1 - x_0) = 1.5 - \frac{-2.25}{11.25} \times 1.5 = 1.8.$$

Hence $F'(x_2) = -1.08$. The next iteration gives

$$x_3 = x_1 - \frac{F'(x_1)}{F'(x_2) - F'(x_1)}(x_2 - x_1) = 3 - \frac{9}{-10.08} \times (-1.2) = 1.9286.$$

The iterates appear to be moving towards the solution $x^* = 2$.

The algorithm we have just used generates each new solution estimate from formula (2.3.2) based upon the two most recently calculated points. In fact, this may not be the most efficient way to proceed. When $k > 1$, we would normally calculate x_{k+2} using x_{k+1} together with either x_k or x_{k-1} according to one of a number of possible strategies:

(a) Choose whichever of x_k and x_{k-1} gives the smaller value of $|F'|$.
(b) Choose whichever of x_k and x_{k-1} gives F' with opposite sign to $F'(x_{k+1})$.
(c) Choose whichever of x_k and x_{k-1} gives the smaller value of F.

Strategies (a) and (c) are based on using points which seem closer to the minimum; strategy (b) seeks to exploit the fact that interpolation is more reliable than extrapolation. Strategy (b), however, can only be employed if we have chosen our initial x_0 and x_1 so that $F'(x_0)$ and $F'(x_1)$ have opposite signs.

To demonstrate strategy (a) we return to the function $F(x) = x^3 - 3x^2$ with the initial points $x_0 = 1.5$ and $x_1 = 3$. As in the worked example above, the first secant iteration gives $x_2 = 1.8$ and so $F'(x_2) = -1.08$. We now need to consider which of x_0 and x_1 should be combined with x_2 in the formula (2.3.2) on the next iteration. Because $|F'(x_0)| = 2.25 < |F'(x_1)|$ we conclude that x_0 is closer to the minimum and so we reassign $x_1 = x_0 = 1.5$. Thus the next iteration gives

$$x_3 = x_1 - \frac{F'(x_1)}{F'(x_2) - F'(x_1)}(x_2 - x_1) = 1.5 - \frac{-2.25}{-1.17} \times 0.3 \approx 2.077.$$

Strategy (a) gives a solution estimate x_3 which is different from the one obtained with the first version of the secant method. The reader can perform further steps to confirm that subsequent x_k converge to the solution $x^* = 2$.

Exercises

1. Apply the secant method to $F(x) = e^x - 2x$ in the range $0 \leq x \leq 1$.
2. Show that (2.3.1) will give $F'(x) = 0$ when applied to any quadratic function $F(x) = ax^2 + bx + c$.
3. Use the secant method with strategy *(b)* on $F(x) = x^3 - 3x^2$ with $x_0 = 1.5$ and $x_1 = 3$. What happens if the starting values are $x_0 = 0.5$ and $x_1 = 1.5$?
4. Suppose that $x_k < x_{k+1}$ and $F'(x_k) > F'(x_{k+1})$. Use a sketch to show that the secant method will give x_{k+2} as a point which approximates a maximum.

2.4.   The Newton method

This method seeks the minimum of $F(x)$ using both first and second derivatives. In its simplest form it can be described as follows.

Newton method for minimizing $F(x)$

Choose x_0 as an estimate of the minimum of $F(x)$
Repeat for $k = 0, 1, 2, \ldots$

$$x_{k+1} = x_k - \frac{F'(x_k)}{F''(x_k)} \qquad (2.4.1)$$

until $|F'(x_{k+1})|$ is sufficiently small.

This algorithm is derived by expanding $F(x)$ as a Taylor series about x_k

$$F(x_k + h) = F(x_k) + hF'(x_k) + \frac{h^2}{2}F''(x_k) + O(h^3). \qquad (2.4.2)$$

Differentiation with respect to h gives a Taylor series for $F'(x)$

$$F'(x_k + h) = F'(x_k) + hF''(x_k) + O(h^2). \qquad (2.4.3)$$

Suppose h is the step from x_k to the minimum x^* so that $F(x_k + h) = 0$. If we assume that h is small enough for the $O(h^2)$ term to be neglected then (2.4.3) implies $h = -F'(x_k)/F''(x_k)$, as used in (2.4.1).

Geometrically, the algorithm can be viewed as using the tangent to the curve of $F'(x)$ to predict where $F'(x)$ itself becomes zero (see Figure 2.2).

As an illustration, we apply the Newton method to $F(x) = x^3 - 3x^2$ for which $F'(x) = 3x^2 - 6x$ and $F''(x) = 6x - 6$. At the initial guess $x_0 = 3$, $F' = 9$ and $F'' = 12$ and so the next iterate is given by

$$x_1 = 3 - \frac{9}{12} = 2.25.$$

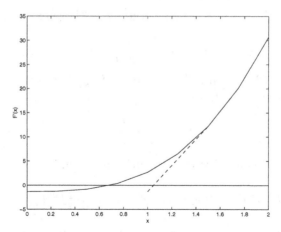

Figure 2.2. Tangent approximation to $F'(x)$ used in the Newton method.

Iteration two uses $F'(2.25) = 1.6875$ and $F''(2.25) = 7.5$ to give

$$x_2 = 2.25 - \frac{1.6875}{7.5} = 2.025.$$

After one more iteration $x_3 \approx 2.0003$ and so Newton's method is converging to the solution $x^* = 2$ more quickly than either bisection or the secant method.

Convergence of the Newton method

Because the Newton iteration is important in the development of optimization methods we study its convergence more formally. We define

$$e_k = x^* - x_k \qquad (2.4.4)$$

as the error in the approximate minimum after k iterations.

Proposition Suppose the Newton iteration (2.4.1) converges to x^*, a local minimum of $F(x)$ where $F''(x^*) = m > 0$. Suppose also there is some neighbourhood N of x^* in which the third derivatives of F are bounded, so that, for some $M > 0$,

$$M \geq F'''(x) \geq -M \quad \text{for all } x \in N. \qquad (2.4.5)$$

If e_k is defined by (2.4.4) then there exists an integer K such that, for all $k > K$,

$$\frac{e_k^2 M}{m} > e_{k+1} > -\frac{e_k^2 M}{m}. \qquad (2.4.6)$$

Proof Because the iterates x_k converge to x^* there exists an integer K such that

$$x_k \in N \quad \text{and} \quad |e_k| < \frac{m}{2M} \quad \text{for } k > K.$$

Then the bounds (2.4.5) on F''' imply $m + M|e_k| > F''(x_k) > m - M|e_k|$. Combining this with the bound on $|e_k|$, we get

$$F''(x_k) > \frac{m}{2}. \tag{2.4.7}$$

Now, by the mean value form of Taylor's theorem,

$$F'(x^*) = F'(x_k) + e_k F''(x_k) + \frac{1}{2} e_k^2 F'''(\xi),$$

for some ξ between x^* and x_k. Because $F'(x^*) = 0$ we deduce

$$F'(x_k) = -e_k F''(x_k) + \frac{1}{2} e_k^2 F'''(\xi).$$

The next estimate of the minimum is $x_{k+1} = x_k - \delta x_k$ where

$$\delta x_k = \frac{F'(x_k)}{F''(x_k)} = -e_k + \frac{e_k^2 F'''(\xi)}{2F''(x_k)}.$$

Hence the error after $k + 1$ iterations is

$$e_{k+1} = x^* - x_{k+1} = e_k + \delta x_k = \frac{e_k^2 F'''(\xi)}{2F''(x_k)}.$$

Thus (2.4.6) follows, using (2.4.5) and (2.4.7).

This result shows that, when x_k is near to x^*, the error e_{k+1} is proportional to e_k^2 and so the Newton method ultimately approaches the minimum very rapidly.

Definition If, for some constant C, the errors e_k, e_{k+1} on successive steps of an iterative method satisfy

$$|e_{k+1}| \leq C e_k^2 \quad \text{as} \quad k \to \infty$$

then the iteration is said to have a *quadratic* rate of ultimate convergence.

Implementation of the Newton method

The convergence result leading to (2.4.6) depends on certain assumptions about higher derivatives and this should warn us that the Newton iteration (2.4.1) may not always be successful. For instance, the calculation will break down if the iterations reach a point where $F''(x)$ is zero.

It is not only this extreme case which causes difficulties, as the following examples show.

Consider $F(x) = x^3 - 3x^2$, and suppose the Newton iteration is started from $x_0 = 1.1$. Because $F'(x) = 3x^2 - 6x$ and $F''(x) = 6x - 6$, we get

$$x_1 = 1.1 - \frac{(-2.97)}{0.6} = 6.05.$$

The minimum of $x^3 - 3x^2$ is at $x = 2$ and so we see that the method has overshot the minimum and given x_1 further away from the solution than x_0.

Suppose now that the Newton iteration is applied to $x^3 - 3x^2$ starting from $x_0 = 0.9$. The new estimate of the minimum turns out to be

$$x_1 = 0.9 - \frac{(-1.89)}{(-0.6)} = -2.25,$$

and the direction of the Newton step is away from the minimum. The iteration is being attracted to the maximum of $F(x)$ at $x = 0$ (which is not unreasonable because the Newton method solves $F'(x) = 0$.)

These two examples show that convergence of the basic Newton iteration depends on the behaviour of $F''(x)$. A practical algorithm should include safeguards against divergence. Clearly we should only use (2.4.1) if $F''(x)$ is strictly positive. We should also check that the new point produced by the Newton formula is "better" than the one it replaces. These ideas are included in the following algorithm which applies the Newton method within a range $[a, b]$ such as can be found by the bracketing algorithm in Section 2.2.

Safeguarded Newton method for minimizing $F(x)$ in $[a, b]$

Make a guess x_0 $(a < x_0 < b)$ for the minimum of $F(x)$
Repeat for $k = 0, 1, 2, \ldots$
if $F''(x_k) > 0$ then
    $\delta x = -F'(x_k)/F''(x_k)$
else
    $\delta x = -F'(x_k)$
if $\delta x < 0$ then $\alpha = \min(1, (a - x_k)/\delta x)$
if $\delta x > 0$ then $\alpha = \min(1, (b - x_k)/\delta x)$
Repeat for $j = 0, 1, \ldots$.
    $\alpha = 0.5^j \alpha$
until $F(x_k + \alpha \delta x) < F(x_k)$
Set $x_{k+1} = x_k + \alpha \delta x$
until $|F'(x_{k+1})|$ is sufficiently small.

As well as giving an alternative choice of δx when $F'' \le 0$, the safeguarded Newton algorithm includes a stepsize α. This is chosen first to prevent the correction steps from going outside the bracket $[a, b]$ and then, by repeated halving, to ensure that each new point has a lower value of F than the previous one. The algorithm always tries the full step ($\alpha = 1$) first and hence it can have the same fast ultimate convergence as the basic Newton method.

We can show the working of the safeguarded Newton algorithm on the function $F(x) = x^3 - 3x^2$ in the range [1,4] with $x_0 = 1.1$. Because

$$F(1.1) = -2.299, \quad F'(1.1) = -2.97 \quad \text{and} \quad F''(1.1) = 0.6$$

the first iteration gives $\delta x = 4.95$. The full step, $\alpha = 1$, gives $x_k + \alpha \delta x = 6.05$ which is outside the range we are considering and so we must reset

$$\alpha = \frac{(4 - 1.1)}{4.95} \approx 0.5859.$$

However, $F(4) = 16 > F(1.1)$ and α is reduced again (to about 0.293) so that

$$x_k + \alpha \delta x = 1.1 + 0.293 \times 4.95 \approx 2.55.$$

Now $F(2.55) \approx -2.93$ which is less than $F(1.1)$. Therefore the inner loop of the algorithm is complete and the next iteration can begin.

Under certain assumptions, we can show that the inner loop of the safeguarded Newton algorithm will always terminate and hence that the safeguarded Newton method will converge.

The difference between the original Newton algorithm and the safeguarded version is worth noting. In most practical optimization algorithms, a simple basic idea has to be augmented by extra features in order to prevent failure when the assumptions behind the method are not satisfied.

Exercises

1. Use Newton's method to estimate the minimum of $e^x - 2x$ in $0 \le x \le 1$. Compare the rate of convergence with that of the bisection method.

2. Show that, for any starting guess, the basic Newton algorithm converges in one step when applied to a quadratic function.

3. Do one iteration of the basic Newton method on the function $F(x) = x^3 - 3x^2$ starting from each of the three initial guesses: $x_0 = 2.1, x_0 = 1, x_0 = -1$. Explain what happens in each case.

4. Do two iterations of the safeguarded Newton method applied to the function $x^3 - 3x^2$ and starting from $x_0 = 0.9$.

5. Devise a safeguarded version of the secant method which restricts the search to a given range $a \leq x \leq b$ and forces it only to accept a new point if it produces a decrease in function value.

Methods using quadratic or cubic interpolation

Each iteration of Newton's method generates x_{k+1} as a stationary point of the interpolating quadratic function defined by the values of $F(x_k)$, $F'(x_k)$ and $F''(x_k)$. In a similar way, a direct-search iterative approach can be based on locating the minimum of the quadratic defined by values of F at three points x_k, x_{k-1}, x_{k-2}; and a gradient approach could minimize the local quadratic approximation given by $F(x_{k-1})$, $F'(x_{k-1})$ and $F(x_k)$. If a quadratically predicted minimum x_{k+1} is found to be "close enough" to x^* (e.g., because $F'(x_{k+1}) \approx 0$) then the iteration terminates; otherwise x_{k+1} is used instead of one of the current points to generate a new quadratic model and hence to predict a new minimum.

As with the Newton method, the practical implementation of this basic idea requires certain safeguards, mostly for dealing with cases where the interpolated quadratic has negative curvature and therefore does not have a minimum. The bracketing algorithm given earlier may prove useful in locating a group of points which implies a suitable quadratic model.

A similar approach is based on repeated location of the minimum of a cubic polynomial fitted either to values of F at four points or to values of F and F' at two points. This method can give faster convergence, but it also requires fall-back options to avoid the search being attracted to a maximum rather than a minimum of the interpolating polynomial.

Exercises

1. Suppose that $F(x)$ is a quadratic function and that, for any two points x_a, x_b, the ratio D is defined by

$$D = \frac{F(x_b) - F(x_a)}{(x_b - x_a)F'(x_a)}.$$

Show that $D = 0.5$ when x_b is the minimum of $F(x)$. What is the expression for D if $F(x)$ is a cubic function?

2. Explain why the secant method can be viewed as being equivalent to quadratic interpolation for the function $F(x)$.

3. Design an algorithm for minimizing $F(x)$ by quadratic interpolation based on function values only.

2.5.   Sample applications and results

We illustrate the performance of the bisection, secant and Newton methods by quoting results obtained using a fortran90 module called OPTIMA. This can be downloaded from an ftp site (as described at the end of the book). It is not essential for the reader to use or understand this software because it should be possible to obtain similar results from other implementations of the methods. (Readers are, in fact, encouraged to program and run their own versions of the minimization algorithms given in this book since this is a very good way to appreciate the advantages and the drawbacks of a particular method.)

Throughout this book we use solutions obtained with the OPTIMA software in order to give a general indication of the relative merits of a number of optimization methods. It should be understood, however, that two implementations of the same method made by different authors will probably not behave in an identical fashion. This is partly because most algorithms involve some arbitrary parameters. For instance, our version of the safeguarded Newton method uses repeated step-halving in the inner iterations to ensure a decrease in the objective function. However, a factor of 0.9 or 0.1 could just as well have been used instead of 0.5. A scaling factor of 0.1 would probably mean that fewer trial steps would be needed on each inner iteration but might also cause the outer iterations to make smaller steps. Other, more subtle, reasons why two implementations of a method may perform differently are considered in later chapters.

The OPTIMA version of the secant method is implemented using strategy *(b)* from Section 2.3. This requires the method to be started with a range $a \leq x \leq b$ which brackets a minimum. The search then maintains a bracket around the solution on subsequent iterations. The Newton method in OPTIMA is an implementation of the safeguarded form of the algorithm. When a search range $a \leq x \leq b$ is specified, the first Newton iteration is started from the midpoint $x = \frac{1}{2}(a + b)$.

Before quoting results, we consider the question of obtaining the derivatives of $F(x)$ which are required by the secant or Newton methods. Sometimes the task of differentiating $F(x)$ will be straightforward, as in the tank design problem. In other cases, such as the total least squares example, the derivatives require more care and effort. It is possible to avoid the work of obtaining analytical expressions for derivatives by using approximations based on finite differencing. Thus we can estimate the slope of the function $F(x)$ at a point $x = a$ by using

$$F'(a) \approx \frac{F(a + h) - F(a)}{h} \qquad (2.5.1)$$

where h is a small positive stepsize. The errors in this approximation can be shown to tend to zero as $h \to 0$. For a given value of h, a more accurate approximation is

$$F'(a) \approx \frac{F(a+h) - F(a-h)}{2h}. \tag{2.5.2}$$

A formula for approximating the second derivative of $F(x)$ at $x = a$ is

$$F''(a) \approx \frac{F(a+h) - 2F(a) + F(a-h)}{h^2}. \tag{2.5.3}$$

Alternatively, if we have an analytical expression for the first derivative $F'(a)$, we can estimate the second derivative from

$$F''(a) \approx \frac{F'(a+h) - F'(a-h)}{2h}. \tag{2.5.4}$$

We make use of finite difference estimates in some examples below.

The tank design problem

In order to apply the secant and Newton methods to the function (1.1.5) we need expressions for the first and second derivatives. It is easy to obtain

$$\frac{dS}{dx_2} = -4V^* x_2^{-2} + 2x_2; \qquad \frac{d^2 S}{dx_2^2} = 8V^* x_2^{-3} + 2.$$

These expressions are used in the program TD0 which lets a user apply the bisection, secant and Newton methods to minimize (1.1.5) for any choice of V^* and from any starting guess for the variable x_2. (Even though the minimum of (1.1.5) can be found by the formula $x_2 = (2V^*)^{1/3}$, we can use this problem to illustrate convergence behaviour of the different methods.)

Taking $V^* = 20$ we consider three possible starting ranges for x_2. (It may not be easy to make a well-informed initial estimate of a solution and a good optimization method should be able to converge from starting points that are badly chosen.) Table 2.1 shows the numbers of iterations needed to find the optimum value $x_2 \approx 3.42$ correct to three decimal places.

Clearly the Newton method is consistently the best approach on this example. The bisection method is relatively inefficient, even when given a fairly good initial guess. However, the bisection method's performance does not deteriorate very much as the search range gets wider whereas the secant method is quite adversely affected.

Starting Range	Bisection Method	Secant Method	Newton Method
$3 \le x_2 \le 4$	14	6	2
$3 \le x_2 \le 5$	15	6	3
$2 \le x_2 \le 6$	16	18	3

Table 2.1. Numbers of iterations to minimize (1.1.5).

The secant method is less efficient than bisection in the third case because the function (1.1.5) has much steeper slopes on the left of the optimum than on the right. By working through the first few iterations, the reader can verify that this causes the left end point to remain unchanged and so convergence is from the right-hand side only. Hence we do not get a bracket which shrinks onto the minimum from both sides. This a fairly common failing of the secant method.

Exercise
For the case in row three of Table 2.1, perform the first three iterations of the secant method (with strategy *(b)* from Section 2.3) and comment on the results. What happens if we use strategies *(a)* or *(c)* instead?

Data-fitting by total least squares

To apply the secant and Newton methods to the function (1.2.4) we need to form its first and second derivatives w.r.t. a. Because $\hat{F}$ depends on a through the intermediate functions τ_i and ϕ in (1.2.3) and (1.2.5), we need a systematic way of organizing the differentiation. If we start with τ_i we can write

$$\tau_i' = \frac{d\tau_i}{da} = \frac{\theta_i}{a^2 + 1} - 2a\frac{t_i + a\theta_i}{(a^2 + 1)^2}$$

$$\tau_i'' = \frac{d^2\tau_i}{da^2} = -2a\frac{\theta_i}{(a^2 + 1)^2} - \frac{2t_i + 4a\theta_i}{(a^2 + 1)^2} + 4a^2\frac{t_i + a\theta_i}{(a^2 + 1)^3}.$$

Then, proceeding to the expressions for ϕ_i, $i = 1, \ldots, 4$,

$$\phi_i' = \frac{d\phi_i}{da} = -\tau_i', \quad \phi_i'' = \frac{d^2\phi_i}{da^2} = -\tau_i''.$$

Also for ϕ_{i+4}, $i = 1, \ldots, 4$,

$$\phi_{i+4}' = \frac{d\phi_{i+4}}{da} = -a\tau_i' - \tau_i, \quad \phi_{i+4}'' = \frac{d^2\phi_{i+4}}{da^2} = -a\tau_i'' - 2\tau_i'.$$

Finally, when $\hat{F}(a)$ is given by (1.2.4),

$$\frac{d\hat{F}}{da} = 2\sum_{i=1}^{4}[\phi_i\phi_i' + \phi_{i+4}\phi_{i+4}']$$

$$\frac{d^2\hat{F}}{da^2} = 2\sum_{i=1}^{4}[(\phi_i')^2 + \phi_i\phi_i'' + (\phi_{i+4}')^2 + \phi_{i+4}\phi_{i+4}''].$$

The demonstration program TLS0 uses these expressions to construct derivatives of the error function (1.2.4), using data from Table 1.1. Table 2.2 shows numbers of iterations needed by the methods to minimize $\hat{F}(a)$ in different search ranges (the optimum value for a is about 2.4046). The relative performance of the methods is similar to that for the tank design problem. The Newton method again does very well whereas the secant method is better than the bisection method only when one end of the search range is very close to the optimum.

Search Range	Bisection Method	Secant Method	Newton Method
$2 \le a \le 3$	14	13	3
$2 \le a \le 5$	15	15	3
$1 \le a \le 6$	16	92	3
$2.4 \le a \le 6$	16	3	4

Table 2.2. Numbers of iterations to minimize (1.2.4).

In order to understand the slow convergence of the secant method in the range $1 \le a \le 6$, we consider Figure 2.3 which shows $\hat{F}(a)$ quite far from the solution.

$\hat{F}(a)$ has negative curvature for values of a greater than about 3.5. Hence the slope is decreasing as a increases towards infinity. This, coupled with the fact that the left-hand end point has a large slope, means that each secant iteration makes a fairly small improvement to the right-hand bracket point. (The reader can verify this by doing a few iterations by hand.)

It is worth noting that the safeguarded Newton method is able to detect negative curvature and to calculate steps that move towards the minimum. The bisection method is unaffected by the negative curvature.

Using approximate derivatives

The program TLS0 can also use numerically estimated derivatives based on (2.5.2) and (2.5.3). Such approximations are often satisfactory and

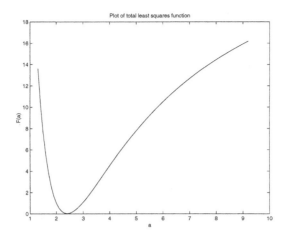

Figure 2.3. Negative curvature in the total least squares error function.

enable a minimization method to locate a solution with (almost) as much accuracy as is possible when analytical derivatives are used. Table 2.3 shows how the differencing stepsize h affects the solutions when the search range is $2 \leq a \leq 3$.

	Secant Method		Newton Method	
h	a^*	itns	a^*	itns
0.1	2.4082	13	2.4082	7
0.01	2.4047	13	2.4047	3
0.001	2.4046	13	2.4046	3

Table 2.3. Minimizing (1.2.4) using numerical derivatives.

The number of iterations does not seem to depend very strongly on h. (The exception is the Newton method when $h = 0.1$ where the number of iterations increases, because each new solution estimate is adversely affected by inaccuracies in both first and second derivatives.) However, the accuracy of the approximate derivatives does influence the quality of the computed solution. The true slope of the best straight-line approximation to the data is $a^* = 2.4046$, correct to five significant figures. As h increases from 0.001, the errors in the estimated derivatives cause the iterations to terminate at points which get further away from the exact minimum.

Exercises

1. Determine exactly the value of a at which $\hat{F}(a)$ begins to have negative curvature.

2. For the case in row three of Table 2.2, perform the first three iterations of the secant method (with strategy *(b)* from Section 2.3) and comment on the results. What happens if we use strategies *(a)* or *(c)* instead?

3. Consider the data points

$$(1, 2.3), \quad (2, 5.1), \quad (3, 7.2), \quad (4, 4.6)$$

and (by using TLS0 or otherwise) determine the straight lines produced by the vertical and total least squares approaches. Comment on the differences between them.

The routing problem

The program R0 solves the problem described in Section 1.3. The results quoted below were obtained using finite difference approximations to derivatives. (But the reader is invited to work out expressions for the analytical derivatives of the functions (1.3.5) and (1.3.7).)

Using the smooth objective function (1.3.7)

We consider the case where the circular obstacle has centre (2,2) and radius 1 and the target point is (5,4). We first minimize the smooth objective function (1.3.7) for various values of ρ and various choices for the search range. The differencing stepsize for derivatives is $h = 0.001$. The minimum of (1.3.7) depends on the value of ρ. Specifically, when $\rho = 1$, $x^* = 1.6178$; when $\rho = 2$, $x^* = 1.6186$; and when $\rho = 4$, $x^* = 1.6190$. When $\rho = 1$, roughly 1% of the optimum route is inside the no-go region. This incursion is approximately halved as ρ is doubled.

Performance of the minimization methods is shown in Tables 2.4–2.6 and we can see that the secant method performs quite poorly. Moreover it tends to do worse as the penalty factor ρ increases, whereas the bisection and Newton methods are relatively unaffected.

Search Range	Bisection Method	Secant Method	Newton Method
$1 \leq x \leq 2$	14	14	10
$0 \leq x \leq 2$	15	12	7
$0 \leq x \leq 3$	15	14	10
$1 \leq x \leq 5$	16	26	11

Table 2.4. Numbers of iterations to minimize (1.3.7) with $\rho = 1$.

Search Range	Bisection Method	Secant Method	Newton Method
$1 \leq x \leq 2$	14	21	8
$0 \leq x \leq 2$	15	21	7
$0 \leq x \leq 3$	15	24	8
$1 \leq x \leq 5$	16	46	5

Table 2.5. Numbers of iterations to minimize (1.3.7) with $\rho = 2$.

Search Range	Bisection Method	Secant Method	Newton Method
$1 \leq x \leq 2$	14	32	8
$0 \leq x \leq 2$	15	26	9
$0 \leq x \leq 3$	15	23	8
$1 \leq x \leq 5$	16	80	6

Table 2.6. Numbers of iterations to minimize (1.3.7) with $\rho = 4$.

Using the nonsmooth objective function (1.3.5)

Strictly speaking, we should only attempt to minimize (1.3.5) with the bisection method, because this is a nonsmooth objective function whose derivatives are not well defined at its minimum. However, we might expect the finite difference formulae to give smoothed approximations to the discontinuous derivatives which will allow us to use the secant and Newton methods after all. To investigate this, we minimize (1.3.5) with $\rho = 1$ using the search range $0 \leq x \leq 2$. The bisection method converges to $x \approx 1.6188$ in 15 iterations. The secant method uses 35 iterations and terminates at a less accurate estimate $x \approx 1.62$. The Newton method stops after 50 iterations at $x \approx 1.6197$ because errors in the approximate derivatives prevent it from meeting the convergence tests. These results show that the use of finite differences may not be a reliable way of attempting to deal with a nondifferentiable function.

We now consider a second routing problem with the no-go region having centre (4,2) and radius 2. The target point is (4,8). Figure 2.4 is a plot of the function (1.3.5) for this problem when $\rho = 1$.

If we apply the bisection method to (1.3.5) in the search range $4 \leq x \leq 10$ we find the optimum turning point is at $x \approx 6.83$ giving an objective function value $F \approx 15.3$. However, if we extend the search range to $0 \leq x \leq 10$ the bisection method finds an optimum at $x = 0$ with $F \approx 8.94$. We can see these two local minima at $x = 0$ and $x \approx 6.83$ in Figure 2.4. The two corresponding locally optimum routes are shown in Figure 2.5. The better of the two (i.e., the global optimum) is OP which goes straight from the origin to the target without even coming

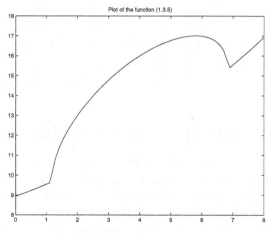

Figure 2.4. Plot of the function (1.3.5).

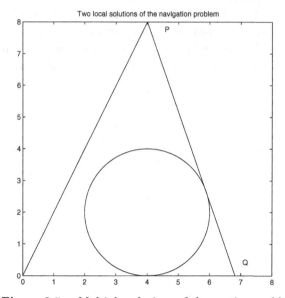

Figure 2.5. Multiple solutions of the routing problem.

near the obstacle. However, the route OQP is the best among routes which pass the obstacle and then turn back. It is locally optimal because a small move of the point Q either to the right or the left would result in a larger value of (1.3.5).

Exercises

1. Calculate an expression for the first derivatives of the function (1.3.7).

2. Solve the routing problem with target point (4,8) and no-go region centred on (4,2) with radius 2 by applying the bisection method to (1.3.7) with $\rho = 1$. Are there multiple local minima for this problem?

3. If the no-go region has centre (4,2) and radius 2 and the target point is (5.4), minimize (1.3.5) with $\rho = 1, 2$ and 4. Comment on the solutions.

Summary of experience with one-variable problems

Results obtained for the previous three example problems confirm that the Newton method can be very efficient in favourable circumstances. The bisection method, on the other hand, is fairly slow but is quite reliable. The secant method shows the biggest variations in performance which suggests that the OPTIMA implementation needs more safeguards.

In all cases, rapid convergence of a method can be seen to be dependent on the initial point or search range being chosen in a region near a solution in which the objective function has a second derivative which is positive (and preferably bounded away from zero).

We show, in the chapters which follow, that one-variable minimization has an important part to play in the solution of problems in n variables.

Chapter 3

Applications in n Variables

The tank design problem introduced in Chapter 1 can involve either
two or three variables (see (1.1.1) and (1.1.2)). The other problems
discussed in Chapter 1 can also be extended to feature more variables.
Thus we might wish to fit a two-parameter model $\theta = at + b$ to the
data in Table 1.1 or to deal with a routing problem involving two or
more turning points. We discuss these cases and some other example
problems in the next sections.

3.1. Data-fitting problems

Consider the problem of fitting a straight line $z = x_1 + x_2 t$ to m data
points $(t_1,\ z_1), \ldots, (t_m,\ z_m)$ using the method of vertical least squares.
We need to find values for the unknown coefficients x_1, x_2 to minimize

$$F(x) = \sum_{i=1}^{m} (z_i - x_1 - x_2 t_i)^2. \tag{3.1.1}$$

It is well known (but we discuss this more formally in the next chapter)
that the first partial derivatives are zero at the minimum of a differentiable function $F(x)$. The reader should verify that

$$\frac{\partial F}{\partial x_1} = -2 \sum_{i=1}^{m} (z_i - x_1 - x_2 t_i) \quad \text{and} \quad \frac{\partial F}{\partial x_2} = -2 \sum_{i=1}^{m} (z_i - x_1 - x_2 t_i) t_i.$$

Setting both expressions to zero implies that the optimum x_1 and x_2
satisfy

$$a_{11} x_1 + a_{12} x_2 = b_1 \quad \text{and} \quad a_{21} x_1 + a_{22} x_2 = b_2 \tag{3.1.2}$$

M. Bartholomew-Biggs, *Nonlinear Optimization with Engineering Applications*,
DOI: 10.1007/978-0-387-78723-7_3, © Springer Science+Business Media, LLC 2008

where

$$a_{11} = m, \quad a_{12} = a_{21} = \sum_{i=1}^{m} t_i, \quad a_{22} = \sum_{i=1}^{m} t_i^2,$$

$$b_1 = \sum_{i=1}^{m} z_i, \quad b_2 = \sum_{i=1}^{m} z_i t_i.$$

Notice that (3.1.2) is a pair of linear equations from which it is easy to obtain x_1 and x_2. At the minimum of any quadratic function the variables will satisfy a system of linear equations and hence it is usual to regard the minimization of a quadratic function as an "easy" problem.

The reader can verify that least squares approximation of m data points by an n-th degree polynomial

$$z = \phi(t, x) = x_1 + x_2 t + x_3 t^2 + \cdots + x_{n+1} t^n$$

also leads to the minimization of a quadratic function which involves the solution of $n + 1$ simultaneous linear equations. However, the approximation of data points by a non-polynomial model usually leads to a nonquadratic objective function. So, for example, if we consider an exponential model

$$z = \phi(t, x) = x_1 e^{x_2 t} \tag{3.1.3}$$

then we need to minimize the function

$$F(x) = \sum_{i=1}^{m} (z_i - x_1 e^{x_2 t})^2. \tag{3.1.4}$$

If we calculate first partial derivatives and set them to zero then the optimal values of x_1 and x_2 satisfy the equations

$$\sum_{i=1}^{m} (z_i - x_1 e^{x_2 t_i}) e^{x_2 t_i} = 0, \quad \sum_{i=1}^{m} (z_i - x_1 e^{x_2 t_i}) e^{x_2 t_i} x_1 t_i = 0$$

which are clearly nonlinear in x_1 and x_2 and not particularly easy to solve. In practice, the minimization of a nonquadratic function such as (3.1.4) is normally done by iterative methods as described in the chapters which follow.

We now consider the total least squares approach to data-fitting when the model function involves two or more parameters. Recall, from Section 1.2, that we require the parameters x_i of the model function to minimize

$$\sum_{i=1}^{m} \hat{r}(t_i, z_i, x)^2 \tag{3.1.5}$$

where $\hat{r}(t_i, z_i, x)$ denotes the shortest distance from point (t_i, z_i) to the curve defined by the model function

$$z = \phi(t, x). \tag{3.1.6}$$

In order to obtain the shortest distance, we first solve the *footpoint problem* to find (t_f, z_f) as the point on the curve (3.1.6) which is closest to (t_i, z_i). This means we obtain t_f by solving the one-variable problem of minimizing

$$(t_i - t)^2 + (z_i - \phi(t, x))^2 \tag{3.1.7}$$

with respect to t. We then obtain

$$\hat{r}(t_i, x_i, x)^2 = (t_i - t_f)^2 + (z_i - \phi(t_f, x))^2. \tag{3.1.8}$$

We now have an interesting situation where the function (3.1.5) involves the subfunctions (3.1.8) which depend on the optimization variables x_i both explicitly and also implicitly through the value of t_f which minimizes (3.1.7).

In the simple case when $\phi(t, x) = x_1 + x_2 t$, (3.1.7) becomes

$$(t_i - t)^2 + (z_i - x_1 - x_2 t)^2$$

which is minimized when its first derivative is zero, which occurs when

$$(t_i - t) + (z_i - x_1 - x_2 t)x_2 = 0.$$

The solution of this equation is

$$t = t_f = \frac{t_i + z_i x_2 - x_1 x_2}{1 + x_2^2}.$$

Then (3.1.8) gives

$$\hat{r}(t_i, x_i, x)^2 = (t_i - t_f)^2 + (z_i - x_1 - x_2 t_f)^2.$$

A more general study of the footpoint problem can be found in [27].

For future reference we define some data-fitting problems to be used in later chapters. In each case we suppose that m data points (t_i, z_i) are to be approximated by a model function $\phi(t, x)$ which involves parameters $x_1, \ldots, x_n$.

Problem VLS uses vertical least squares and seeks the minimum of

$$F(x) = \sum_{i=1}^{m} (z_i - \phi(t_i, x))^2. \tag{3.1.9}$$

Problem TLS uses total least squares and seeks the minimum of

$$F(x) = \sum_{i=1}^{m}(t_i - t_f)^2 + (z_i - \phi(t_f, x))^2 \qquad (3.1.10)$$

where t_f minimizes (3.1.7).

Exercises

1. Show that the function (3.1.9) is quadratic when $\phi(t, x)$ is the nth degree polynomial $z = x_1 + x_2t + x_3t^2 + \cdots + x_{n+1}t^n$.
2. A solution to problem VLS when $\phi(t, x) = x_1e^{x_2t}$ can be approximated by fitting the model $\log_e z = \log_e x_1 + x_2t$ to the data using vertical least squares. Show that this can be posed as a quadratic minimization problem.
3. Write down expressions for the first partial derivatives of (3.1.9) and (3.1.10) in terms of partial derivatives of ϕ.
4. A way of estimating t_f to minimize (3.1.7) is to use simple iteration to seek a point which makes the first derivative of (3.1.7) zero. We want t_f to solve

$$(t_i - t) + (z_i - \phi(t, x))\frac{d\phi(t, x)}{dt} = 0$$

and we can use the iterative scheme

$$t_f^{(k+1)} = t_i + (z_i - \phi(t_f^{(k)}, x))\frac{d\phi(t_f^{(k)}, x)}{dt}, \qquad \text{for } k = 0, 1, 2, \ldots$$

with the initial guess $t_f^{(0)} = t_i$. Try this approach to formulate the objective function (3.1.10) when the model function is given by (3.1.3).

3.2.   The routing problem

The routing problem in section 1.3 can be extended if the vehicle's initial movement is not forced to be along the x-axis. Suppose instead that it can move to any point (x_1, y_1) before turning towards the target (x_p, y_p). The vehicle may now enter the no-go region during both stages of the route and so we consider the general problem of determining the intersection between the circle and a line segment from (x_b, y_b) to (x_e, y_e). Any point on this segment can be written as $(x_b + \lambda(x_e - x_b), y_b + \lambda(y_e - y_b))$ with $0 \leq \lambda \leq 1$. Points of intersection with the circle occur when

$$(x_b + \lambda(x_e - x_b) - x_c)^2 + (y_b + \lambda(y_e - y_b) - y_c)^2 = r^2.$$

This implies λ satisfies $a\lambda^2 + b\lambda + c = 0$ where

$$a = (x_e - x_b)^2 + (y_e - y_b)^2,$$

$$b = 2((x_b - x_c)(x_e - x_b) + (y_b - y_c)(y_e - y_b))$$

$$c = (x_b - x_c)^2 + (y_b - y_c)^2 - r^2.$$

If this equation has complex roots then there is no intersection between the segment and the circle. If, however, there are real roots λ_1, λ_2 we have to consider whether the intersections lie between $(x_b, \ y_b)$ and $(x_e, \ y_e)$. We can assume, without loss of generality, that the roots are numbered so that $\lambda_1 \leq \lambda_2$. If we write the total segment length as

$$d(x_b, y_b, x_e, y_e) = \sqrt{(x_e - x_b)^2 + (y_e - y_b)^2}$$

then $\nu(x_b, y_b, x_e, y_e)$, the segment length inside the circle, is found as follows

$$\text{if }\ 0 \leq \lambda_1 \leq \lambda_2 \leq 1\ \text{ then }\ \nu = (\lambda_2 - \lambda_1)d(x_b, y_b, x_e, y_e)$$

$$\text{if }\ 0 \leq \lambda_1 \leq 1\ \text{ and }\ \lambda_2 > 1\ \text{ then }\ \nu = (1 - \lambda_1)d(x_b, y_b, x_e, y_e)$$

$$\text{if }\ \lambda_1 < 0 \text{ and }\ 0 \leq \lambda_2 \leq 1\ \text{ then }\ \nu = \lambda_2 d(x_b, y_b, x_e, y_e)$$

$$\text{if }\ \lambda_1 < 0\ \text{ and }\ \lambda_2 > 1\ \text{ then }\ \nu = d(x_b, y_b, x_e, y_e)$$

$$\text{if }\ \lambda_1 > 1\ \text{ or }\ \lambda_2 < 0\ \text{ then }\ \nu = 0.$$

Using this notation, the optimum route can be found by minimizing the smooth objective function

$$d(0, 0, x_1, y_1) + d(x_1, y_1, x_p, y_p) + \rho[\nu(0, 0, x_1, y_1)^3 + \nu(x_1, y_1, x_p, y_p)^3].$$
$$(3.2.1)$$

For future reference we call this *Problem R1(1)*.

Exercises

1. Construct an expression for the objective function of *Problem R1(2)* which allows the vehicle to make two turns between the origin and the target point.
2. Construct a revised version of Problem R1(1) for the case when the obstacle is described by an ellipse $\alpha(x - x_c)^2 + \beta(y - y_c)^2 = 1$.
3. Derive a three-dimensional version of Problem R1(1) for the case when the obstacle is defined by the ellipsoid

$$(x - x_c)^2 + (y - y_c)^2 + \gamma\ z^2 = r^2.$$

3.3.   An optimal control problem

In this section we use a simple model of the motion of a train to determine an operating policy for the driver when accelerating or decelerating. We represent the vehicle as a body moving in a straight line. At an initial time $(t = 0)$ its distance from a reference origin is s_0 and its speed is u_0. We consider the train's position and speed at n equally-spaced times $t = 0, \tau, 2\tau, \ldots, (n-1)\tau$ and we suppose that x_k denotes the constant applied acceleration between times $(k-1)\tau$ and $k\tau$. If u_k, s_k denote the body's speed and distance from the origin at time $t = k\tau$ then, for $k = 1, \ldots, n$,

$$u_k = u_{k-1} + x_k \tau \quad \text{and} \quad s_k = s_{k-1} + u_{k-1}\tau + \frac{1}{2}x_k \tau^2. \qquad (3.3.1)$$

We want to choose $x_1, \ldots, x_n$ to make s_n and u_n as close as possible to some given values s_f and u_f and so we want to minimize

$$(s_n - s_f)^2 + (u_n - u_f)^2.$$

This is not the whole story, however, because the operation of a passenger vehicle should not involve accelerations that are large or rapidly changing. Bearing this in mind, we consider the minimization of the function

$$F(x_1, \ldots, x_n) = (s_n - s_f)^2 + (u_n - u_f)^2 + \rho P \qquad (3.3.2)$$

where s_n and u_n are given by (3.3.1) and

$$P = x_1^2 + x_n^2 + \sum_{k=2}^{n} (x_k - x_{k-1})^2. \qquad (3.3.3)$$

By including P in $F(x)$ we are involving a "smoothness" requirement in the objective function. The parameter ρ is a weighting factor which reflects the importance given to smoothness of the motion. We refer to this as *Problem OC1(n)* where n denotes the number of time steps. Larger values of n correspond to more accurate models of the original problem in which time is continuously varying.

Problem OC2(n) imposes the smoothness condition in a slightly different way, by defining

$$P = x_1^2 + x_n^2 + \sum_{k=2}^{n} \left(1 - \frac{x_k}{x_{k-1}}\right)^2. \qquad (3.3.4)$$

This measures relative differences between the x_k whereas (3.3.3) measures absolute differences.

Examples such as OC1 and OC2 are typical of a large class of practical problems involving the optimization of some dynamic system whose continuous behaviour is approximated by discretization over a number of small time steps. Such problems are interesting because they enable us to observe how optimization methods behave as the number of variables becomes large.

Exercises

1. Derive expressions for

$$\frac{\partial F}{\partial x_i} \quad \text{and} \quad \frac{\partial^2 F}{\partial x_i^2}$$

   when F is given by (3.3.2) and (3.3.3).
2. Repeat Question 1 when F is defined by (3.3.2) and (3.3.4).
3. Solve the minimization problem defined by (3.3.2) and (3.3.3) when $n = 2$ and $s_0 = u_0 = 0$, $s_f = 1$, $u_f = 0$, and $\tau = 0.5$.

Core curriculum [3]

Frowning at the inkwell of learning
tongue thrust out through lips
labouring over letters
with a nib that splits
to spatter extra dots on all the i's
and cross the t's before we come to them.
Each character perfected
to help us form strong characters
we rehearse an alphabet
for spelling out our stories.

1950's child and this years adult:
both striving for connections
to give our words the virtue of integrity
and – something that we learned about much later –
more value than their letters' algebraic sum.
There's still some doubt that either one will manage
to get to grips with proper joined-up writing.

Chapter 4

n-Variable Unconstrained Optimization

4.1.   Optimality conditions

Definition Suppose $F(x)$ is a continuous function of x, where $x = (x_1, \ldots, x_n)^T$. If, at some point $x = x^*$, there exists $\epsilon > 0$ such that

$$F(x^*) \le F(x) \quad \text{when} \quad ||x - x^*|| \le \epsilon \qquad (4.1.1)$$

then $F(x)$ is said to have a *local minimum* at x^*.

When $F(x)$ is an n-variable continuously differentiable function, the conditions which characterise a minimum can be expressed in terms of the vector of first partial derivatives

$$g = \left(\frac{\partial F}{\partial x_1}, \ldots, \frac{\partial F}{\partial x_n} \right)^T, \qquad (4.1.2)$$

and the $n \times n$ matrix G of second partial derivatives whose (i,j)th element is

$$G_{ij} = \frac{\partial^2 F}{\partial x_i \partial x_j}. \qquad (4.1.3)$$

Definitions The vector g in (4.1.2) is called the *gradient* and may also be written as ∇F (or sometimes as F_x). The matrix G given by (4.1.3) is known as the *Hessian* and may also be denoted by $\nabla^2 F$ or F_{xx}.

The Hessian matrix is always symmetric when F is a twice continuously differentiable function because then

$$G_{ij} = \frac{\partial^2 F}{\partial x_i \partial x_j} = \frac{\partial^2 F}{\partial x_j \partial x_i} = G_{ji}.$$

This is the case for most of the problems we consider.

M. Bartholomew-Biggs, *Nonlinear Optimization with Engineering Applications*,
DOI: 10.1007/978-0-387-78723-7_4, © Springer Science+Business Media, LLC 2008

Definition A *positive-definite* symmetric matrix is one which has all positive eigenvalues. Equivalently, a matrix A is positive-definite if and only if

$$x^T A x > 0, \quad \text{for any } x \neq 0. \tag{4.1.4}$$

Definition If $F(x)$ is an n-variable function whose gradient and Hessian satisfy

$$g(x^*) = 0 \quad \text{and} \quad G(x^*) \text{ is positive-definite.} \tag{4.1.5}$$

then the point x^* is a *local minimum* of $F(x)$.

It is the second of the *optimality conditions* (4.1.5) that distinguishes a minimum from a maximum (or any other stationary point) because it ensures $F(x^*) < F(x)$ for all x in some, possibly small, region around x^*. For some functions $F(x)$ there may be several points x^* which satisfy (4.1.5). These are all *local* minima; and the one which gives the least value of F is called the *global* minimum.

From a geometrical point of view, the positive-definiteness of G implies that the function is *convex* near the minimum. (Convexity is briefly discussed in Section 4.2.)

If a point $\bar{x}$ has $g(\bar{x}) = 0$ but $G(\bar{x})$ *negative-definite* (i.e., has all negative eigenvalues) then $\bar{x}$ is a local maximum of $F(x)$.

If $\bar{x}$ is such that $g(\bar{x}) = 0$ and $G(\bar{x})$ is *indefinite* (i.e., has both positive and negative eigenvalues) then $\bar{x}$ is a *saddle point*. This means that $\bar{x}$ appears to be a minimum when approached along some directions but resembles a maximum when approached along others. To visualise this, consider the surface defined by the function $F(x) = x_1 x_2$ which has a saddle point at the origin.

If $g(\bar{x}) = 0$ and $G(\bar{x})$ is positive *semi-definite* (i.e., $v^T G(\bar{x}) v \geq 0$ because $G(\bar{x})$ has a zero eigenvalue) then it is necessary to consider third derivatives of $F(x)$ to determine whether $\bar{x}$ is a weak local minimum or a saddle point.

We can sometimes use (4.1.5) directly to minimize $F(x)$. Consider the problem

$$\text{Minimize } F(x_1, x_2) = (x_1 - 1)^2 + x_2^3 - x_1 x_2. \tag{4.1.6}$$

Setting the first partial derivatives to zero gives

$$2x_1 - 2 - x_2 = 0 \quad \text{and} \quad 3x_2^2 - x_1 = 0.$$

These equations have two solutions

$$(x_1, x_2) = \left(\frac{3}{4}, -\frac{1}{2} \right) \quad \text{and} \quad (x_1, x_2) = \left(\frac{4}{3}, \frac{2}{3} \right).$$

To identify the minimum we consider the Hessian

$$G = \begin{pmatrix} 2 & -1 \\ -1 & 6x_2 \end{pmatrix}.$$

We can show quite easily that G is positive definite when $x_2 = \frac{2}{3}$ but not when $x_2 = -\frac{1}{2}$. Hence the minimum is at $(\frac{4}{3}, \frac{2}{3})$.

In practice we cannot tackle problems in this way when the first-order condition $g = 0$ yields equations which cannot be solved analytically. For such problems we must use iterative methods. Many of these are gradient techniques which – like the secant or Newton methods – require the calculation of first (and sometimes second) derivatives. However some n-dimensional minimization methods are like the one-variable bisection technique and use only function values. These are called *direct search* methods.

Exercises

1. Prove that, when conditions (4.1.5) hold, $F(x^* + sp) > F(x^*)$ for any vector p, provided the scalar s is sufficiently small.
2. Use the optimality conditions to minimize (1.1.2) when $V^* = 20$.
3. The eigenvalues of a matrix G can be found by solving the characteristic equation $det(G - \lambda I) = 0$. Use this to show that the Hessian matrix of (4.1.6) is positive-definite when x_2 is positive.

4.2.   Visualising problems in several variables

It is easy to illustrate one-variable optimization problems using graphs of the objective function. For two-dimensional problems we can use *contour plots* and these can also give some insight into problems in higher dimensions. A contour plot for a function of two variables shows curves in the (x_1, x_2)-plane along which the function has a constant value (as on maps which show lines of constant altitude.) Figure 4.1 is a contour plot of a function whose minimum is near the middle of the central oval region. The closed curves are contours whose function value increases towards the edges of the figure.

Definition   *Convexity* of a function means that any two points lying inside one of its contour lines can be joined by a straight line which also lies entirely inside the same contour line. Equivalently, $F(x)$ is convex if and only if $F(a) < \hat{F}$ and $F(b) < \hat{F}$ implies $F(a + \lambda(b - a)) < \hat{F}$ for $0 \le \lambda \le 1$.

The contours in Figure 4.1 show that the function is convex, in the range of x_1, x_2 values illustrated. Pictorially, a nonconvex function has

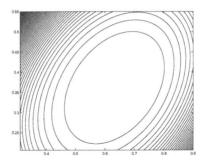

Figure 4.1.   Contours of a convex function.

contour lines which "double-back on themselves" as in Figures 4.2 and
4.3. If a function $F(x)$ is convex for all x then it has a unique minimum.
Nonconvex functions, however, may have multiple stationary points. In
Figure 4.3 there is a maximum near $x_1 = 0.5$, $x_2 = 0.33$ and a saddle
point near $x_1 = 0.42$, $x_2 = 0.27$.

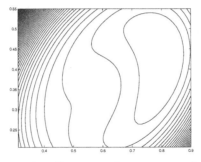

Figure 4.2.   Contours of a nonconvex function.

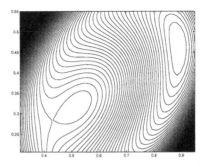

Figure 4.3.   Multiple stationary points of a nonconvex function.

4.3.   Optimization software and test problems

We have already mentioned the OPTIMA software in Chapter 2. OPTIMA is written in fortran90 and includes implementations of most of the optimization techniques described in the chapters which follow. OPTIMA can be downloaded from an ftp site (as described at the end of the book) and used to solve many of the examples and exercises given in subsequent chapters.

One way of comparing the performance of two iterative minimization methods is to consider the numbers of iterations they take to solve the same set of problems. (We must, of course, use the same starting guesses for the variables and the same convergence test.) However, an iteration count is not the only measure of efficiency. The amount of computing effort needed to solve an optimization problem also depends on the number of evaluations of the objective function (and perhaps its derivatives). Thus, when we quote results obtained with OPTIMA, we typically state both the number of iterations performed and also the number of calls to the procedure which calculates $F(x)$ (together with the gradient and Hessian if these are used).

It should be emphasised that use of OPTIMA is not essential to the understanding of this book. The sample problems can be handled by other optimization software such as SOLVER [29] in Microsoft Excel [48], the MATLAB optimization toolbox [61] or codes from the optimization chapter of the NAG library [62]. The results obtained with any implementation of a particular method should be broadly similar to those that we report from OPTIMA. It is worth mentioning, however, that the routines in OPTIMA may not be as "highly tuned" for rapid convergence as those in some of the commercial packages we have mentioned. As we explain more fully in later chapters, the performance of a numerical algorithm depends not only on the soundness of the underlying theory but also on the details of its implementation.

The convergence tests used in OPTIMA can be briefly described as follows. We have already distinguished between direct search algorithms which only evaluate the objective function $F(x)$ and gradient techniques which use derivatives of F. The direct search routines in OPTIMA stop iterating when a point x is found such that

$$|F(x) - F(x^-)| < \epsilon(\epsilon + |F(x)|) \quad \text{or} \quad ||x - x^-||_2 < \epsilon\sqrt{n} \quad (4.3.1)$$

where x^- denotes the solution estimate obtained on a previous iteration. The gradient-based routines in OPTIMA employ a convergence test of the form

$$||\nabla F(x)||_2 < \epsilon\sqrt{n} \quad (4.3.2)$$

for terminating the iterations. In both (4.3.1) and (4.3.2) the standard accuracy criterion uses $\epsilon = 10^{-5}$. The OPTIMA software also allows us to choose low- or high-accuracy solutions corresponding, respectively, to $\epsilon = 10^{-4}$ and $\epsilon = 10^{-6}$.

We now give details of some test problems whose solutions are reported in the chapters which follow. These problems have been implemented as Fortran programs which can be downloaded along with OPTIMA.

Problem TD1 is the minimization of (1.1.2) with $V^* = 20$ starting from the initial guess $x_1 = x_2 = 2$. This problem has solution $x_1^* \approx 1.71$, $x_2^* \approx 3.42$ with a minimum surface area ≈ 35.09 (see the contour plots in Figure 4.4).

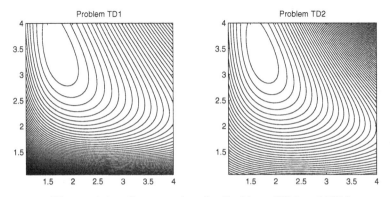

Figure 4.4.   Contour plots for Problems TD1 and TD2.

Problem TD2 is the minimization of

$$V = -\frac{x_1 x_2 (S^* - 2x_1 x_2)}{2x_1 + x_2} \qquad (4.3.3)$$

with $S^* = 35$ and starting from $x_1 = x_2 = 2$. (This is a formulation of the problem of maximizing tank volume subject to a limit on surface area.) TD2 has solution $x_1^* \approx 1.708$, $x_2^* \approx 3.416$ with maximum volume ≈ 19.92. We may suspect (correctly) that the solutions of Problems TD1 and TD2 would be identical if the value of S^* in TD2 were 35.09. We say there is a *dual* relationship between the problems: if S^* is the minimum surface area for a given volume V^* then V^* is the maximum volume for a given surface area S^*.

It is important to remember that Problems TD1 and TD2 involve functions which are convex only in the neighbourhood of the minimum. Both (1.1.2) and (4.3.3) tend to infinity as x_1 and x_2 approach zero; and they take large negative values if both x_1 and x_2 become negative. It is only meaningful to search for a minimum when x_1 and x_2 are positive.

Problem VLS1 determines the best approximation (using vertical least squares) of the data points $(t_i, \ z_i) = (0,3), \ (1,8), \ (2,12), \ (3,17)$ by minimizing (3.1.9) when ϕ is the straight line $z = x_1 + x_2 t$. The starting guess is $x_1 = x_2 = 0$. The solution is $x_1^* = 3.1, x_2^* = 4.6$.

Problem TLS1 uses total least squares to approximate the points $(0,3)$, $(1,8)$, $(2,12)$, $(3,17)$ by minimizing (3.1.10) when ϕ is the straight line $z = x_1 + x_2 t$. The starting guess is $x_1 = x_2 = 0$. The solution is $x_1 = 3.0875$, $x_2 = 4.6083$. This is slightly different from that given by applying vertical least squares to the problem. The contours shown in Figure 4.5 are perfect ellipses because the objective function for Problem VLS1 is quadratic. The contours for TLS1 (Figure 4.6) may appear to be elliptical, but in fact they are not and the nonquadratic nature of the objective function makes problem TLS1 harder to solve, as shown by the results quoted in later chapters.

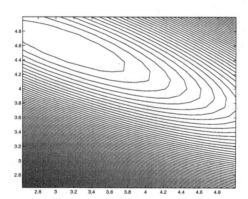

Figure 4.5. Contour plot for Problem VLS1.

Problem VLS2 determines the vertical least squares approximation to the points

$$(t_i, \ z_i) = (0,1), \ (1,0.5), (2,0.4), \ (3,0.3), \ (4,0.2)$$

by minimizing (3.1.9) when ϕ is the exponential function $z = x_1 e^{x_2 t}$. The starting guess is $x_1 = x_2 = 1$. The solution is $x_1 \approx 0.9515$, $x_2 \approx -0.4434$ giving the sum of squared errors as about 0.0185. As there are five data points, the *root mean square* error is given by

$$F_{rms} \approx \sqrt{\frac{1}{5}(0.0185)} \approx 0.061$$

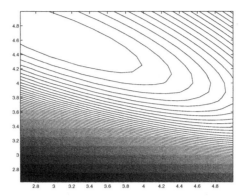

Figure 4.6.   Contour plot for Problem TLS1.

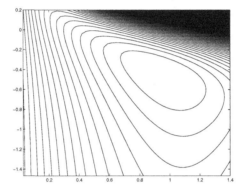

Figure 4.7.   Contour plot for Problem VLS2.

which can be taken as a rough estimate of the average residual at each data point. Contours of the objective function appear in Figure 4.7 which makes clear the nonquadratic nature of the problem.

Problem R1(1) solves the problem described in Section 3.2 and determines the optimum turning point for a two-stage route from $(0,0)$ to $(8,4)$ when the no-go circle is defined by $x_c = 4$, $y_c = 3$, $r = 2$. The objective function (3.2.1) uses the weighting parameter $\rho = 0.1$. The starting guess is the point $(4,2)$ and the solution is $x^* \approx 4.976$, $y^* \approx 1.177$ and $F^* \approx 9.251$. The contours of (3.2.1) around the solution can be seen in Figure 4.8. They clearly show the nonconvexity of this problem. Figure 4.9 displays the contours of (3.2.1) over a wider range of values for the optimization variables. This shows that there is a second, locally optimal, route with $x^* \approx 3.029$, $y^* \approx 5.311$ giving an objective function value $F^* \approx 11.272$. This is clearly inferior to the solution already quoted above, but it represents the best route among all those which pass above the no-go region.

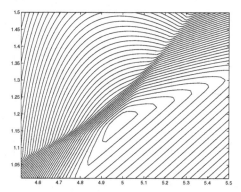

Figure 4.8. Near-optimum contour plot for Problem R1(1).

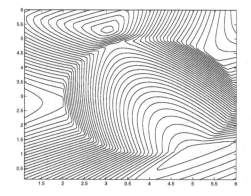

Figure 4.9. Contour plot for Problem R1(1).

Problem R1(2) seeks the optimum 3-stage route with the same data as Problem R1(1). The starting guess has turning points (3.9, 1.95) and (4.9, 2.45). The exact solution makes turns at (4.712, 1.113) and (5.176, 1.362) giving an optimum function value $F^* \approx 9.233$.

We could also pose a Problem R1(m) involving m turning points. But in fact there is nothing practical to be gained by making more than two turns to get around the circular obstacle. Figure 4.10 shows solutions obtained for R1(m) as m increases. The upper subgraphs show results when $m = 1$ and $m = 2$ and the introduction of the second turning point produces a small decrease in the objective function value from about 9.25 to about 9.23. However, there is no further reduction in the objective function when $m = 3$ or 4. The lower subgraphs show that the extra "turning points" are simply placed on the last segment of the route and have no effect either on the overall route length or on the extent of penetration of the no-go circle. In other words the variables

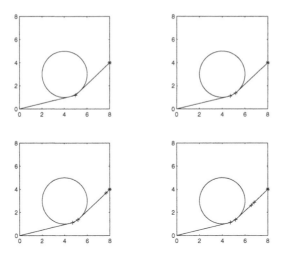

Figure 4.10.   Solutions of Problem R1(m) for $m = 1, 2, 3, 4$.

which define the third and subsequent turning points are redundant in a problem concerned with optimal routing around a single circle.

The reader can consult [65, 9] for extensions of the above ideas to deal with more than one obstacle.

Problem OC1(n) is the optimal control problem from Section 3.3 which minimizes (3.3.2) with $\rho = 0.01$ and P defined by (3.3.3). The initial and final conditions are defined by

$$t_f = 3, \quad \tau = \frac{t_f}{n}, \quad s_0 = 0, \quad s_f = 1.5, \quad u_0 = 0, \quad u_f = 0$$

and so the problem is to determine a smooth acceleration profile which will take the vehicle from rest to rest over a distance of 1.5 km in 3 minutes. If there are only two time steps then the symmetrical solution is to accelerate with $x_1 = \frac{2}{3}$ km/min^2 and then decelerate with $x_2 = -\frac{2}{3}$ km/min^2. If there are n time steps (n is assumed to be even) then the starting guess is

$$x_i = 0.66, \quad i = 1, \ldots, \frac{n}{2}; \quad x_i = -0.66, \quad i = \frac{n}{2} + 1, \ldots, n.$$

Figure 4.11 shows the solutions of Problem OC1(n) as n increases. The pattern that emerges is one where the acceleration increases to a maximum and then decreases until braking starts about halfway through the motion. The braking force then mirrors the acceleration with maximum braking being applied about three-quarters of the way through the motion.

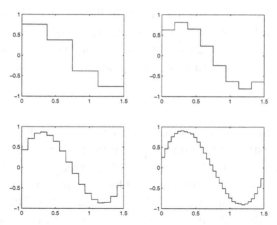

Figure 4.11. Solutions of Problem OC1(n) for $n = 4, 8, 16, 32$.

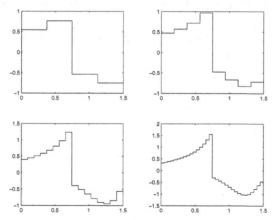

Figure 4.12. Solutions of Problem OC2(n) for $n = 4, 8, 16, 32$.

Problem OC2(n) is the same as Problem OC1(n) except that P is defined by (3.3.4). Figure 4.12 shows the solutions of this problem as n increases.

In contrast to the results for Problem OC1, we see that the acceleration increases until around the halfway point. Braking then is applied quite sharply, after which the deceleration history is somewhat similar to that for Problem OC1. Clearly, the relative measure of smoothness in (3.3.4) does not produce the required result midway through the journey when x_i and x_{i-1} have opposite signs. If x_{i-1} is positive and x_i is negative then, in order to make $1 - x_i/x_{i-1}$ small, the optimization forces x_{i-1} to be large while decreasing the magnitude of x_i. This is what we see in the calculated solutions of OC2 and it indicates that the optimization model in this problem is not a good one it presents interesting. In spite of this, however, we use OC2 as one of our examples because it presents interesting challenges to some of the optimization methods studied in later chapters.

Exercises

1. Prove the dual relationship between Problems TD1 and TD2 mentioned in the paragraph following Equation (4.3.3).

2. Obtain a formulation of a problem TLS2 which computes an exponential fit to the data points for Problem VLS2 using the total least squares approach. This will involve minimizing the function (3.1.10) after using an iterative method to find t_f to minimize (3.1.7).

3. Obtain a formulation for extended versions of Problems R1(1) and R1(2) in which there are two circular obstacles to be avoided. How would these problems be further extended to allow more than two turning points?

4. Consider the objective function for Problem OC2(2) and plot its contours for $-1 \leq x_1,\ x_2 \leq 1$.

Chapter 5

Direct Search Methods

This book is mainly concerned with gradient methods of minimization; but in this chapter we consider approaches which resemble the bisection method in relying only on function values. These are usually referred to as *direct search* methods and are useful when we want to minimize functions which are not (or not easily) differentiable.

Obviously we could search for the least value of a function by evaluating it at all points on a "grid" of values of the variables; but clearly this is not very efficient. Alternatively, we might approximate the minimum more rapidly by sampling the function value at a sequence of "random" points, using statistical arguments to estimate the likelihood of finding the minimum in a certain number of trials. Both approaches are sometimes used; but we confine ourselves to techniques that are more systematic.

5.1.   Univariate search

Univariate search is based on performing a sequence of one-dimensional minimizations of $F(x)$, first with respect to x_1, then with respect to x_2 and so on. In other words, we search for a minimum point along each of the coordinate directions in turn. We can avoid gradient calculations by using a direct-search method such as bisection. At the end of a complete "cycle" of minimizations, all n variables will have been adjusted; but many such cycles will usually be needed to locate the overall optimum x^*. This method sometimes works quite well (e.g., on two-variable functions with near-circular contours) but, in general, it is not guaranteed to converge.

M. Bartholomew-Biggs, *Nonlinear Optimization with Engineering Applications*,
DOI: 10.1007/978-0-387-78723-7_5, © Springer Science+Business Media, LLC 2008

As an example we apply the method to the function

$$F(x) = x_1^2 + 2x_1x_2 + 3x_2^2 \qquad (5.1.1)$$

starting from the initial guess $x_1 = x_2 = 1$. The algorithm first explores the x_1 direction to minimize $x_1^2 + 2x_1 + 3$. Hence (e.g., by using bisection in the range $0 \leq x_1 \leq 2$) it obtains $x_1 = -1$. The method then searches the x_2 direction for the minimum of $1 - 2x_2 + 3x_2^2$ and (via bisection in $0 \leq x_2 \leq 1$) obtains $x_2 = \frac{1}{3}$.

Hence the first cycle yields the point $x = (-1, \frac{1}{3})^T$ where $F = \frac{2}{3}$ which is much smaller than the function value $F = 6$ at the starting point $(1, 1)^T$.

The reader can verify that the second cycle of one-dimensional searches yields the solution estimate $x = (\frac{1}{3}, -\frac{1}{9})$ (which is closer to $x^* = (0, 0)^T$).

The algorithm is spelled out below. We assume the minimization of $F(x)$ is confined to a "hyperbox" $l_i \leq x_i \leq u_i$, $i = 1, \ldots, n$. To distinguish iteration numbers from subscripted elements in a vector we let x_{k_i} be the ith variable after k iterations. We use e_i to denote the ith column of the $n \times n$ identity matrix.

Univariate search for minimizing Fx) in $l \leq x \leq u$

Choose an initial estimate x_0 of the minimum of $F(x)$
Repeat for $k = 0, 1, 2, \ldots$
Set $x^+ = x_k$
Repeat for $i = 1, \ldots, n$
  Find s^* to minimize $F(x^+ + se_i)$ in range $u_i - x_i^+ \geq s \geq x_i^+ - l_i$.
  Set $x^+ = x^+ + s^*e_i$
Set $x_{k+1} = x^+$
until $\|x_{k+1} - x_k\|$ is sufficiently small

An extension of univariate search is called the *Hooke and Jeeves method*. This technique [37] augments each cycle of one-dimensional minimizations with a "pattern move" which entails a further line search along the resultant direction obtained by adding the individual one-variable moves. Thus the Hooke and Jeeves algorithm is the same as univariate search except that at the end of the kth iteration there is an extra step involved in obtaining x_{k+1}:

Find s^* to minimize $F(x^+ + s(x^+ - x_k))$
Set $x_{k+1} = x^+ + s^*(x^+ - x_k)$

We can illustrate the use of a pattern move on the function (5.1.1). The calculations above show that the first cycle of univariate search makes a step from $x^{(0)} = (1, 1)^T$ to a new point $x^+ = (-1, \frac{1}{3})^T$ Hence the pattern move gives

$$x^{(1)} = x^{(0)} + s^*(x^+ - x^{(0)}) = \left(1 - 2s^*, \; 1 - \frac{2s^*}{3}\right)^T$$

where s^* is chosen to minimize

$$(1 - 2s)^2 + 2(1 - 2s)\left(1 - \frac{2s}{3}\right) + 3\left(1 - \frac{2s}{3}\right)^2.$$

By applying the bisection method in the range $0 \le s \le 1$ we get $s^* = \frac{5}{8}$ and so the point reached by the first Hooke and Jeeves iteration is

$$x^{(1)} = \left(-\frac{1}{4}, \; \frac{7}{12}\right)^T.$$

This is closer to the optimum $x^* = (0, 0)^T$ than the point $(-1, \frac{1}{3})^T$ given by the univariate search on its own.

Exercises

1. Do two cycles of univariate minimization on the function (4.1.6) starting from $x_1 = x_2 = 1$. What happens if a pattern move is added?
2. Continue the worked example above and perform a second iteration of the univariate search method on problem (5.1.1) to show that it yields the new point $x = (-\frac{1}{3}, \frac{1}{9})^T$. Show also that a pattern move from this point will locate the exact minimum of (5.1.1).
3. Implement and test a computational procedure for univariate search, using the bisection method as the one-dimensional minimization algorithm.
4. Extend your procedure from the previous question to implement a pattern move as in the Hooke and Jeeves algorithm.

5.2. The Nelder and Mead simplex method

This method [50] (not to be confused with the simplex method in linear programming) is usually more effective than univariate search. We first outline the approach for a two-variable optimization problem.

For a two-variable problem we make three initial estimates of the position of the minimum. These will define a starting *simplex*. (More generally, for a function of n variables, a simplex consists of $n+1$ points.) Suppose we label the *vertices* of the simplex A, B, C and call the corresponding function values F_a, F_b, F_c. The vertex with the highest function value is said to be the worst; and this point must be replaced with a better one.

The basic move in the simplex method is *reflection*. A new trial point is obtained by reflecting the worst point in the centroid of the remaining vertices. This is a heuristic way of placing a new solution estimate in a region where lower function values are likely to occur. Suppose, for instance, that $F_a > F_b > F_c$. Then the vertex A would be reflected in the centroid of vertices B and C. Let the new point be labelled as N and let F_n be the associated function value. If $F_n < F_a$ the new point is an improvement on vertex A and a new simplex is defined by deleting the old worst point and renaming vertex N as A.

We can illustrate the reflection step if the simplex method is applied to (5.1.1) using the initial points A at $(1,0)^T$, B at $(0,1)^T$ and C at $(1,1)^T$. The corresponding function values are, respectively, $F_a = 1$, $F_b = 3$ and $F_c = 6$. Now we reflect the point C in the centroid of the other two which is at $(\frac{1}{2}, \frac{1}{2})^T$. Hence the new point N is

$$\left(1 + 2\left(\frac{1}{2} - 1 \right), \ 1 + 2\left(\frac{1}{2} - 1 \right) \right)^T = (0, \ 0)^T.$$

In this case the reflection step has, rather fortuitously, located the minimum!

To show that a reflection move is not always so successful we consider (5.1.1) when the starting simplex has points A, B and C at $(1,0)^T$, $(0, \frac{1}{2})^T$ and $(1,1)^T$. C is still the point with the highest function value and so we get N by reflecting C in the centroid of A and B at $(\frac{1}{2}, \frac{1}{4})^T$. Hence N is

$$\left(1 + 2\left(\frac{1}{2} - 1 \right), \ 1 + 2\left(\frac{1}{4} - 1 \right) \right)^T = \left(0, -\frac{1}{2} \right)^T.$$

Because $F_n = \frac{3}{4}$, which is less than $F_c = 6$, this new point is acceptable and replaces C as a vertex of the new simplex.

It can sometimes happen that the reflected point is not acceptable. If the initial simplex for problem (5.1.1) has vertices A, B, C at $(1,0)^T$, $(-1,0)^T$, $(0,1)^T$ then $F_a = F_b = 1$ and $F_c = 3$. Hence point C is reflected in the centroid $(0,0)^T$ which gives N at $(-1,0)^T$. But this will mean that $F_n = F_c = 3$ and so N is not a suitable replacement for C. In such a case, the simplex method uses *modified reflection* to generate another trial point M so that $CM = \frac{2}{3}(CN)$. If $F_m < F_c$ then M becomes part of a new simplex. The reader can verify that modified reflection does give an acceptable new vertex in the example we are considering.

If both reflection and modified reflection fail to give an acceptable new point then the minimum may lie inside the simplex. In these circumstances a *contraction* strategy is used. This involves shrinking the current simplex towards its best point. Thus, if B is the lowest vertex, a new simplex is obtained by halving the distances of all other vertices from B. The method stops when the simplex has shrunk below a certain size.

Formalising and extending these ideas for an n-variable problem we get the following algorithm. The reader will note that it is significantly more complicated than other methods we have considered so far. (Correct implementation of such an algorithm is a nontrivial task; but a version that is fairly widely available is the procedure FMINSEARCH in MATLAB [61].)

Nelder and Mead simplex algorithm for minimizing $F(x)$

Choose points $x_0, \ldots, x_n$ to form an initial simplex
Repeat
Find x_w such that $F(x_w) \geq F(x_i)$ for $i = 1, \ldots, n$
Find x_b such that $F(x_b) \leq F(x_i)$ for $i = 1, \ldots, n$
Calculate

$$\tilde{x} = \frac{1}{n-1} \sum_{i=1, i \neq w}^{n} x_i \quad \text{and} \quad \bar{x} = \frac{1}{n} \sum_{i=1}^{n} x_i.$$

Set $x^+ = x_w + 2(\tilde{x} - x_w)$ (reflection)
If $F(x^+) < F(x_b)$ (reflection is successful)
set $x^{++} = x^+ + (x^+ - x_w)$ (expand reflected step)
if $F(x^{++}) < F(x^+)$ then set $x^+ = x^{++}$
set $x_w = x^+$
else
if $F(x^+) < F(x_w)$
set $x_w = x^+$
else   (reflection is unsuccessful)
set $x^+ = x_w + \frac{4}{3}(\tilde{x} - x_w)$ (shrink reflected step)
If $F(x^+) < F(x_w)$
set $x_w = x^+$
else   (modified reflection is unsuccessful)
   for $i = 1, \ldots, n$, $i \neq b$ set $x_i = \frac{1}{2}(x_i + x_b)$ (contract towards x_b)
Until $||x_i - \bar{x}||$ is sufficiently small for $i = 1, \ldots, n$

Exercise
Do two iterations of the simplex method on the function (4.1.6) starting from an initial simplex with corners at $(1, 1)$, $(1.5, 1)$, $(1, 1.5)$.

5.3. DIRECT

DIRECT [39] is a particularly interesting technique because – unlike most of the methods in this book – it seeks the global rather than a local minimum of $F(x)$. In practice, global minimizers are usually applied in some restricted region, typically in a "hyperbox" defined by $l_i \leq x_i \leq u_i$. DIRECT relies on the use of such rectangular bounds and works by systematic exploration of rectangular subregions. In the limit, as the number of iterations becomes infinite, it will sample the whole region and, in that sense, the algorithm is guaranteed to converge. The practical performance of the method depends on how it chooses which subregions to explore first, because this determines whether the global minimum can be approximated in an acceptable number of iterations.

To describe the method, we consider first the one-variable problem of finding the global minimum of $F(x)$ for $0 \leq x \leq 1$. We begin by dividing $[0, 1]$ into three equal subranges and evaluating $F(x)$ at their midpoints. The range containing the least function value is taken to be the "most promising" and so we trisect it, evaluating F at the midpoints of the new ranges. We then have a situation of the kind shown in Figure 5.1.

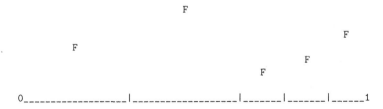

Figure 5.1.   One iteration of DIRECT on a one-variable problem.

There are now trial ranges of two different widths, namely, $\frac{1}{3}$ and $\frac{1}{9}$. For each of these widths we trisect the one with the smallest value of F at the centre. This is depicted in Figure 5.2.

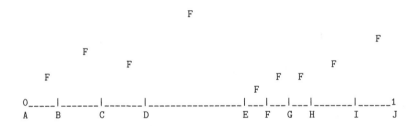

Figure 5.2.   Two iterations of DIRECT on a one-variable problem.

The situation shown in Figure 5.2 involves three candidate range-sizes, $\frac{1}{3}$, $\frac{1}{9}$ and $\frac{1}{27}$. For each of these, the third iteration trisects the one which has the smallest F-value at its centre. In the diagram, this would mean subdividing the intervals DE, AB and EF. Continuing in this way, we can systematically explore the whole range in a way that concentrates on the most promising regions first. Thus we aim to find a good estimate of the global optimum before the iteration count gets too high.

The basic idea just outlined can be made more efficient if we refine the definition of a "promising" range. Let $d_1, \ldots, d_p$ be the p different rangesizes at the start of an iteration and let F_j denote the smallest of all the function values at the centres of ranges of width d_j. The range containing F_j is trisected only if a "potential optimality" test is satisfied. This test is based upon *Lipschitz constants*, which are bounds on the size of the first derivative of F. If F has Lipschitz constant L then, within the range containing F_j, we have the bounds

$$F_j + \frac{1}{2}Ld_j \geq F(x) \geq F_j - \frac{1}{2}Ld_j.$$

We do not normally know a Lipschitz constant for F. However, the range containing F_j can be said to be potentially optimal if there exists a Lipschitz constant L such that

$$F_j - \frac{1}{2}Ld_j < F_i - \frac{1}{2}Ld_i \quad \text{for} \quad i = 1, \ldots, p; \quad i \neq j. \qquad (5.3.1)$$

If (5.3.1) holds then it is possible that the range containing F_j also contains a smaller value of F than can be found in any other range. For (5.3.1) to be satisfied we need

$$L > 2 \times \max \left\{ \frac{F_j - F_i}{d_j - d_i} \right\} \quad \text{for all } i : d_i < d_j$$

and

$$L < 2 \times \min \left\{ \frac{F_j - F_i}{d_i - d_j} \right\} \quad \text{for all } i : d_i > d_j$$

If these conditions on L are inconsistent then the range containing F_j cannot be considered potentially optimal and hence it need not be subdivided. This consideration can save wasteful function evaluations when there are many different candidate ranges.

Another "filter" can be used to reduce the number of ranges to be subdivided. If $F_{\min}$ is the smallest function value found so far then the range containing F_j will not be trisected unless there exists an L which satisfies (5.3.1) and also

$$F_j - \frac{1}{2}Ld_j < F_{\min} - \hat{\epsilon}|F_{\min}| \qquad (5.3.2)$$

where $\hat{\epsilon}$ is a user-specified parameter. Condition (5.3.2) suggests that subdivision of the range containing F_j can be expected to produce a nontrivial improvement in the best function value so far.

The above ideas can be extended to problems in several variables [39]. The original search region becomes a hyperbox rather than a line segment and the initial subdivision is into three hyperboxes by trisection along the longest edge. The objective function is evaluated at the centre point of each of these boxes and the size of each box is taken as the length of its diagonal. The box with the smallest value of F at its centre is subdivided by trisection along its longest side and the process of identification and subdivision of potentially optimal hyperboxes then continues as in the one-variable case. (There are refinements for subdividing boxes with several longest sides [39].)

Experience has shown that DIRECT can often get good estimates of global optima quite quickly [9]. It only uses function values and so it can be applied to nonsmooth problems or to those where the computation of derivatives is difficult. One drawback, however, is that there is no hard-and-fast convergence test for stopping the algorithm. One can simply let it run for a fixed number of iterations or else terminate if there is no improvement in the best function value after a prescribed number of function evaluations. Neither strategy, however, will guarantee to identify the neighbourhood of the global optimum.

5.4.   Results with direct search methods

The results quoted below show the performance of UNIVAR and DIRECT which are the OPTIMA implementations of univariate search and the DIRECT algorithm. We also include results obtained with FMINSEARCH, the MATLAB implementation of the Nelder and Mead simplex algorithm [61]. UNIVAR uses the bisection method for the one-dimensional minimizations. Table 5.1 shows the numbers of iterations and function calls needed by the methods to solve Problems TD1–OC2 to standard accuracy defined by (4.3.1) with $\epsilon = 10^{-5}$. Each iteration of UNIVAR consists of a complete cycle of n one-dimensional searches parallel to each coordinate axis. DIRECT uses the value $\hat{\epsilon} = 0.01$ in the test (5.3.2). Both UNIVAR and DIRECT require a search range to be specified for each x_i. This is used to establish a bracket for each one-dimensional minimization in UNIVAR and to define the initial hyperbox for DIRECT. The search ranges used in the quoted results are:

TD1, TD2: $0.1 \leq x_i \leq 3.9$ for $i = 1, 2$
VLS1, TLS1: $-5 \leq x_i \leq 5$ for $i = 1, 2$

VLS2: $0 \leq x_1 \leq 2, \quad -1 \leq x_2 \leq 3$
R1(1): $3 \leq x_1 \leq 5, \quad 1 \leq y_1 \leq 3$
R1(2):
  $2.9 \leq x_1 \leq 4.9, \quad 0.95 \leq y_1 \leq 2.95, \quad 3.9 \leq x_2 \leq 5.9, \quad 1.25 \leq y_2 \leq 3.25$
OC1(4), OC2(4):
  $0 \leq x_i \leq 1.32$ for $i = 1, 2$ and $-1.32 \leq x_i \leq 0$ for $i = 3, 4$

In each case, the starting values for the variables are given by the mid-point of the stated range.

Problem	UNIVAR itns/fns	FMINSEARCH itns/fns	DIRECT itns/fns
TD1	5/370	37/72	23/241
TD2	5/370	37/72	33/367
VLS1	18/1404	74/145	33/787
TLS1	18/1404	81/152	33/773
VLS2	21/1512	46/86	33/503
R1(1)	29/2030	37/71	33/325
R1(2)	31/4340	199/335	55/1227
OC1(4)	168/22176	119/206	75/2243
OC2(4)	32/4224	97/168	55/1543

Table 5.1.   Direct search solutions for Problems TD1–OC2.

The results in Table 5.1 show the importance of considering numbers of function evaluations as well as numbers of iterations. On iteration count alone, FMINSEARCH usually appears inferior to both UNIVAR and DIRECT; but, in terms of function calls, it is much more efficient. Each simplex iteration can be seen to require only one or two function calls whereas an iteration of UNIVAR involves n accurate minimizations and is much more expensive. On Problem TD1, for instance, each UNIVAR iteration uses about 90 function calls, that is 45 for each bisection search. (This cost might be reduced if the one-dimensional minimizations were performed less accurately.) The DIRECT iterations are cheaper than those of UNIVAR but still require between 8 and 40 function evaluations. Because DIRECT seeks a global minimum, each of its iterations may evaluate the function at points throughout the whole search region whereas FMINSEARCH confines its exploration to a region near the current simplex.

Exercises

1. How would you expect the performance of DIRECT to change if $\hat{\epsilon} > 0.01$? Use numerical tests to see if your expectations are confirmed.

2. What would you expect to happen to the performance of UNIVAR and DIRECT if the search ranges were smaller than those quoted before Table 5.1? Use numerical tests to see if your expectations are confirmed.

3. Implement and test a version of univariate search in which the one-variable bisection method is only required to find a minimum to low accuracy.

4. Apply the direct search methods from this chapter to Problems R1(1) and R1(2) with $\rho > 0.1$. Comment on the solutions and the computational costs of obtaining them.

5. Apply the simplex method and DIRECT to Problem R1(1), using the starting guess $x_1 = 3$, $y_1 = 5$ and the search range $1 \leq x_1 \leq 5$, $1 \leq y_1 \leq 9$ and comment on the results.

Chapter 6

Computing Derivatives

6.1.   Hand-crafted derivatives

In the chapters which follow we mainly deal with optimization methods
which are iterative gradient techniques involving the calculation of first
(and sometimes second) partial derivatives at each solution estimate.
As noted in Chapter 2, the analytic differentiation of objective functions
which occur in practical applications may be a nontrivial process. In
such situations it can be helpful to take a systematic approach. As
an example of the careful *hand-crafting* of derivatives we consider the
differentiation of the function (3.3.2). For notational convenience we use
$s'_{k,i}$, $u'_{k,i}$ to denote $\partial s_k/\partial x_i$ and $\partial u_k/\partial x_i$, respectively. Because s_0 and
u_0 are given we must have

$$s'_{0,i} = \frac{\partial s_0}{\partial x_i} = u'_{0,i} = \frac{\partial u_0}{\partial x_i} = 0 \quad \text{for} \quad i = 1, \ldots, n.$$

Then, using (3.3.1) we get, for $i = 1, \ldots, n$

$$s'_{k,i} = \frac{\partial s_k}{\partial x_i} = \begin{cases} s'_{k-1,i} + \tau u'_{k-1,i} & \text{if } i < k \\ \frac{1}{2}\tau^2 & \text{if } i = k \\ 0 & \text{if } i > k \end{cases}$$

$$u'_{k,i} = \frac{\partial u_k}{\partial x_i} = \begin{cases} u'_{k-1,i} & \text{if } i < k \\ \tau & \text{if } i = k. \\ 0 & \text{if } i > k \end{cases}$$

M. Bartholomew-Biggs, *Nonlinear Optimization with Engineering Applications*,
DOI: 10.1007/978-0-387-78723-7_6, © Springer Science+Business Media, LLC 2008

By using these expressions for $k = 1, \ldots, n$ we can compute all the elements of ∇s_n and ∇u_n, where

$$\nabla s_n = \left(\frac{\partial s_n}{\partial x_1}, \ldots, \frac{\partial s_n}{\partial x_n} \right)^T \quad \text{and} \quad \nabla u_n = \left(\frac{\partial u_n}{\partial x_1}, \ldots, \frac{\partial u_n}{\partial x_n} \right)^T.$$

From (3.3.3) we have

$$\frac{\partial P}{\partial x_1} = 2x_1 - 2(x_2 - x_1); \quad \frac{\partial P}{\partial x_n} = 2x_n + 2(x_n - x_{n-1})$$

and

$$\frac{\partial P}{\partial x_i} = 2(x_i - x_{i-1}) - 2(x_{i+1} - x_i) \quad \text{for} \quad i = 2, \ldots, n-1$$

which give us all the elements of the gradient vector ∇P. Hence we can compute the gradient of the complete objective function (3.3.2) as

$$\nabla F = 2(s_n - s_f)\nabla s_n + 2(u_n - u_f)\nabla u_n + \rho \nabla P. \tag{6.1.1}$$

A formula like (6.1.1) would have to be derived and then coded, along with the expression for the objective function $F(x)$, before a gradient-based optimization routine could be applied.

Exercises
1. Determine expressions for the second partial derivatives of (3.3.2).
2. Determine expressions for ∇F and $\nabla^2 F$ when F is the objective function for Problem OC2(n).

Derivatives in matrix and vector notation

Functions of many variables can sometimes be written in compact form using matrix and vector notation. If v is an n-vector and c is a constant scalar then a general n-variable linear function has the form

$$F = v^T x + c. \tag{6.1.2}$$

It is easy to show (see exercises below) that the gradient and Hessian of (6.1.2) are

$$\nabla F = v, \quad \nabla^2 F = 0. \tag{6.1.3}$$

If M is a symmetric $n \times n$ matrix then an n-variable quadratic function is of the form

$$F = x^T M x + v^T x + c. \tag{6.1.4}$$

The gradient and Hessian of (6.1.4) are given by

$$\nabla F = 2Mx + v, \quad \nabla^2 F = 2M. \tag{6.1.5}$$

Expressions (6.1.3) and (6.1.5) can be regarded as basic identities which are useful when differentiating more complicated expressions. Thus, for instance, we can use the function-of-a-function rule to say that the quadratic function

$$F = (v^T x)^2 \tag{6.1.6}$$

has the gradient vector

$$\nabla F = 2(v^T x)v. \tag{6.1.7}$$

Similarly, the quartic function

$$F = (x^T M x)^2 \tag{6.1.8}$$

has the gradient

$$\nabla F = 4(x^T M x)M x. \tag{6.1.9}$$

By applying the rule for differentiation of a product we can show that the cubic function

$$F = (v^T x)(x^T M x) \tag{6.1.10}$$

has a gradient given by

$$\nabla F = (x^T M x)v + 2(v^T x)M x. \tag{6.1.11}$$

Similarly, we can use the product rule to differentiate (6.1.7) and hence obtain the Hessian of (6.1.6) as

$$\nabla^2 F = 2vv^T. \tag{6.1.12}$$

Further applications of these ideas appear in the exercises below.

Exercises

1. Verify the results (6.1.3), (6.1.5) for the cases $n = 2$ and $n = 3$.
2. If N is a 2×2 nonsymmetric matrix show that the gradient of the function $F = x^T N x$ is $\nabla F = (N + N^T)x$. What is $\nabla^2 F$?
3. If $F = \frac{1}{2} x^T M x + v^T x + c$ find expressions for ∇F and $\nabla^2 F$.
4. Find an expression for the Hessian matrix of (6.1.8) when $n = 2$ and deduce the corresponding result for general n.
5. Use the rule for differentiating a quotient to obtain expressions for ∇F if F is given by

$$\text{(i)} \quad F = \frac{v^T x}{x^T M x} \quad \text{and} \quad \text{(ii)} \quad F = \frac{x^T M x}{v^T x + c}.$$

6. If α, β are scalar constants and $F = (\alpha x + \beta)^T M (\alpha x + \beta)$ find ∇F and $\nabla^2 F$.

6.2.   Finite difference estimates of derivatives

We stated in Section 2.5 that derivatives of a one-variable function $F(x)$ can be approximated by finite difference formulae. We now treat this idea a little more formally and extend it to functions of several variables.

The standard definition of the first derivative of a one-variable function $F(x)$ is

$$\frac{dF}{dx} = \lim_{h \to 0} \frac{F(x+h) - F(x)}{h}.$$

Hence, for any particular value of x, we can approximate the first derivative by choosing a small value for h and setting

$$\frac{dF}{dx} \approx \frac{F(x+h) - F(x)}{h}. \tag{6.2.1}$$

This is called the *forward difference* approximation.

The accuracy of the derivatives estimated by (6.2.1) depends upon h being neither "too big" nor "too small". If we rearrange the Taylor series expansion

$$F(x+h) = F(x) + h\frac{dF(x)}{dx} + \frac{h^2}{2}\frac{d^2F(x)}{dx^2} + \frac{h^3}{6}\frac{d^3F(x)}{dx^3} + O(h^4), \tag{6.2.2}$$

to give

$$\frac{F(x+h) - F(x)}{h} = \frac{dF(x)}{dx} + \frac{h}{2}\frac{d^2F(x)}{dx^2} + \frac{h^2}{6}\frac{d^3F(x)}{dx^3} + O(h^3)$$

then it is clear that (6.2.1) gives an error which is $O(h)$. If the chosen value of h is big enough to make this error significant in comparison with the true first derivative then the approximation is said to be contaminated by *truncation error*. If h is very small, however, the truncation error will also be small but the approximation (6.2.1) can then be damaged by *rounding error*. When h is near zero, the values computed in finite-precision arithmetic for $F(x+h)$ and $F(x)$ may differ in only one or two digits and so the right-hand side of (6.2.1) will give poor accuracy.

As an illustration, consider forward difference estimates of the first derivative of $F(x) = \sqrt{x}$ for different values of x. Calculations in 15-digit real arithmetic give the results in Table 6.1 which show the percentage error in the approximate derivative for different values of x and h.

In each case the errors initially decrease with h and then start to increase again once h becomes too small. However the values of h that are "too big" or "too small" vary with x. Hence the use of finite difference approximations to derivatives may require a preliminary trial-and-error investigation to determine a value for h which ensures that the

h	10^{-1}	10^{-4}	10^{-7}	10^{-10}	10^{-13}
% error($x = 1$)	2.4	2×10^{-3}	2×10^{-6}	8×10^{-6}	8×10^{-2}
% error($x = 100$)	2×10^{-3}	2×10^{-5}	8×10^{-6}	9×10^{-3}	6.6
% error($x = 0.01$)	53	0.25	2×10^{-4}	3×10^{-6}	2×10^{-3}

Table 6.1. Errors in forward differencing.

computed derivatives are not too much damaged by either truncation error or rounding.

This idea behind (6.2.1) extends easily to functions of several variables. For an n-variable function $F(x_1, \ldots, x_n)$ the forward difference approximations to the first partial derivatives are

$$\frac{\partial F}{\partial x_i} \approx \frac{F(x + he_i) - F(x)}{h} \quad \text{for } i = 1, \ldots, n \qquad (6.2.3)$$

where e_i is the ith column of the identity matrix.

An alternative estimate for first derivatives of a one-variable function is

$$\frac{dF}{dx} \approx \frac{F(x + h) - F(x - h)}{2h}. \qquad (6.2.4)$$

This is the *central difference* formula, derived by subtracting the Taylor series

$$F(x - h) = F(x) - h\frac{dF(x)}{dx} + \frac{h^2}{2}\frac{d^2 F(x)}{dx^2} - \frac{h^3}{6}\frac{d^3 F(x)}{dx^3} + O(h^4) \quad (6.2.5)$$

from (6.2.2). The errors in (6.2.4) are $O(h^2)$. Table 6.2 shows that, for a given step size h, the percentage errors in the derivative estimates given by (6.2.4) are usually smaller than those from (6.2.1). Notice, however, that errors still tend to increase as h becomes too large and also as h approaches zero.

h	10^{-1}	10^{-4}	10^{-7}	10^{-10}	10^{-13}
% error($x = 1$)	0.12	1×10^{-7}	6×10^{-8}	8×10^{-6}	3×10^{-2}
% error($x = 100$)	1×10^{-5}	1×10^{-8}	8×10^{-6}	9×10^{-3}	6.6
% error($x = 0.01$)	73	1×10^{-3}	1×10^{-9}	1×10^{-6}	8×10^{-4}

Table 6.2. Errors in central differencing.

Obviously there is a version of the central difference formula for first partial derivatives of a function of n variables, namely

$$\frac{\partial F}{\partial x_i} \approx \frac{F(x + he_i) - F(x - he_i)}{2h} \quad \text{for } i = 1, \ldots, n. \qquad (6.2.6)$$

Finite difference approximations can also be used for second derivatives. If we add (6.2.2) and (6.2.5) the odd-powered terms cancel and we get

$$F(x + h) + F(x - h) = 2F(x) + h^2 \frac{d^2 F(x)}{dx^2} + O(h^4).$$

From this there follows the central difference estimate of the second derivative of a one-variable function $F(x)$,

$$\frac{d^2 F}{dx^2} \approx \frac{F(x + h) - 2F(x) + F(x - h)}{h^2}. \tag{6.2.7}$$

The error in this formula is $O(h^2)$. Analagous ideas can be used to estimate the second derivatives of an n-variable function, so that

$$\frac{\partial^2 F}{\partial x_i^2} \approx \frac{F(x + he_i) - 2F(x) + F(x - he_i)}{h^2} \tag{6.2.8}$$

and

$$\frac{\partial^2 F}{\partial x_i \partial x_j} \approx \frac{F(x + hs_{ij}) - F(x + hd_{ij}) - F(x - hd_{ij}) + F(x - hs_{ij})}{4h^2}$$

$$\tag{6.2.9}$$

where

$$s_{ij} = e_i + e_j \quad \text{and} \quad d_{ij} = e_i - e_j.$$

Finite difference schemes such as these have been widely used in practical optimization. However, there can be difficulties in choosing the stepsize h to ensure that the approximate derivatives are sufficiently accurate. These difficulties may be avoided if we use the techniques described in the next section.

Exercise
Using (6.2.6), (6.2.8) and (6.2.9) calculate estimates of the gradients and Hessians of the functions in Problems TD1 and TD2 when $x_1 = x_2 = 1$.

6.3.   Automatic differentiation

The term *automatic differentiation* (AD) is used to denote computational techniques implemented in software tools which apply the rules for differentiating sums, products, functions of functions and so on. Such tools can be interfaced with a program for calculating a mathematical expression in order to evaluate first (and higher) derivatives along with the expression itself. AD is distinct from – although clearly related to – software for symbolic differentiation which operates on mathematical formulae and produces corresponding formulae for derivatives with respect to chosen variables.

In the context of optimization, automatic differentiation can provide derivatives of the objective function without requiring a user to do any calculus. Essentially it takes a user-supplied program for evaluating $F(x)$ and then carries out extra computations, based on the rules of calculus, to obtain the corresponding derivative value(s). Some software tools do the extra computations at the same time as the function evaluation; but others construct a separate procedure for the derivative calculation. (The first approach is said to use *overloaded operations* while the second is called *preprocessing*.) In both cases the derivatives are evaluated as accurately as the function itself (i.e., subject only to possible rounding errors). Truncation errors do not arise as they do for derivatives estimated by finite differences.

The simplest software tools for automatic differentiation involve the introduction of a new data type and a set of associated operations.

Definition The *doublet* data type is a bracketed pair of the form $U = \{u, u'\}$. Here u is called the *value* and u' the *gradient* of U.

We first consider the use of doublets for differentiating functions of one variable. In this case the value and gradient are both real scalars. If $U = \{u, u'\}$ and $V = \{v, v'\}$ then basic doublet arithmetic operations are

$$U + V = \{u + v, u' + v'\} \quad \text{and} \quad U - V = \{u - v, u' - v'\} \quad (6.3.1)$$

$$UV = \{uv, vu' + uv'\} \quad (6.3.2)$$

$$\frac{U}{V} = \left\{ \frac{u}{v}, \frac{u'v - v'u}{v^2} \right\}. \quad (6.3.3)$$

The gradient parts of (6.3.1)–(6.3.3) capture the rules for differentiating sums, products and quotients. There are similar definitions for mixed operations between doublets and constants. If $U = \{u, u'\}$ and c is a real constant then

$$U + c = \{u + c, u'\} \quad \text{and} \quad U - c = \{u - c, u'\} \quad (6.3.4)$$

$$Uc = cU = \{uc, u'c\} \quad (6.3.5)$$

$$\frac{U}{c} = \left\{ \frac{u}{c}, \frac{u'}{c} \right\} \quad \text{and} \quad \frac{c}{U} = \left\{ \frac{c}{u}, -\frac{cu'}{u^2} \right\}. \quad (6.3.6)$$

We can also extend the meanings of standard functions to allow them to take doublet arguments. If $U = \{u, \ u'\}$ then, for instance,

$$\sin(U) = \{\sin(u), \ u' \cos(u)\} \tag{6.3.7}$$

$$\log(U) = \left\{ \log(u), \ \frac{u'}{u} \right\}. \tag{6.3.8}$$

$$U^n = \{u^n, \ n u^{n-1} u'\} \quad \text{if } n \neq 0 \text{ is an integer.} \tag{6.3.9}$$

In a similar way we can define the doublet extension of any differentiable real-valued unary or binary functions.

Definition If $h(u)$ is a real-valued differentiable function of a real scalar argument u then its *doublet extension* for $U = \{u, \ u'\}$ is

$$H(U) = \{h(u), \ u' h_u\} \quad \text{where} \quad h_u = \frac{dh(u)}{du}. \tag{6.3.10}$$

Definition If $h(u, v)$ is a real-valued differentiable function of two real scalar arguments u and v then its doublet extension for $U = \{u, \ u'\}$, $V = \{v, \ v'\}$ is

$$H(U, V) = \{h(u, v), \ u' h_u + v' h_v\} \quad \text{where} \quad h_u = \frac{\partial h}{\partial u} \quad \text{and} \quad h_v = \frac{\partial h}{\partial v}. \tag{6.3.11}$$

The basic doublet operations (6.3.1)–(6.3.3) are particular cases of (6.3.11) and the mixed operations (6.3.4)–(6.3.6) are particular cases of (6.3.10). Hence (6.3.10) and (6.3.11) effectively sum up the rules of doublet calculation.

If $f(x)$ is a differentiable function of a single variable x we can obtain its first derivative by evaluating f according to the rules of doublet arithmetic. To do this we must first convert the independent variable to doublet form.

Definition If $f(x)$ is a function of a scalar variable x then its *doublet extension* $F(X)$ is obtained by replacing x by its *doublet form*

$$X = \{x, \ 1\}. \tag{6.3.12}$$

The definition (6.3.12) is consistent with the fact that, trivially, x is the value of the variable and $x' = 1$ is the gradient of x with respect to itself.

The evaluation of the doublet extension $F(X)$ uses the rules (6.3.10), (6.3.11). As an example, consider the function

$$f(x) = x^3 \sin x + \cos x^2. \tag{6.3.13}$$

Its doublet extension F is

$$F = X^3 \sin X + \cos X^2 \qquad (6.3.14)$$

where X is given by (6.3.12). Using (6.3.1)–(6.3.11) for any x we get

$$F = \{x^3,\ 3x^2\}\{\sin x,\ \cos x\} + \{\cos x^2,\ -2x \sin x^2\}$$
$$= \{x^3 \sin x,\ 3x^2 \sin x + x^3 \cos x\} + \{\cos x^2,\ -2x \sin x^2\}$$
$$= \{x^3 \sin x + \cos x^2,\ 3x^2 \sin x + x^3 \cos x - 2x \sin x^2\}.$$

The reader can verify that the gradient part of F is what we should have obtained by differentiating (6.3.13) in the usual way. In particular, if $x = 1$, the evaluation of (6.3.14) gives

$$F = \{1,\ 3\}\{0.84147,\ 0.5403\} + \{0.5403,\ -2 \times 0.84147\}$$
$$= \{0.84147,\ 3 \times 0.84147 + 1 \times 0.5403\} + \{0.5403,\ -1.68294\}$$
$$= \{1.38177,\ 1.38177\}.$$

This agrees with the fact that, for the function (6.3.13), both $f(x)$ and its first derivative $f'(x)$ simplify to $\sin 1 + \cos 1$ when $x = 1$.

Generalising the above example, the following result is the basis of the *forward accumulation method* of automatic differentiation.

Proposition If $f(x)$ is a differentiable function of a scalar variable x and $F(X)$ is its doublet extension then the gradient part of $F(X)$ will give $f'(x)$ for any x for which this derivative exists.

The above ideas can be extended to functions of n variables. In this case the value part of a doublet is still a scalar but the gradient part becomes an n-vector. However, all the rules (6.3.1)–(6.3.11) still apply, and only the definition (6.3.12) needs to be modified.

Definition If $f(x)$ is a function of n independent variables $x_1, \ldots, x_n$ then its doublet extension is obtained by replacing each x_i by its *doublet form*

$$X_i = \{x_i,\ e_i\} \quad i = 1, \ldots, n \qquad (6.3.15)$$

where e_i denotes the ith column of the unit matrix.
With this definition we can state the following result.

Proposition If $f(x)$ is a differentiable function of n variables $x_1, \ldots, x_n$ and $F(X)$ is its doublet extension then the gradient part of $F(X)$ will give ∇f for any x for which this gradient exists.

As an example we consider the two variable function

$$f(x) = x_1(x_1 + x_2). \tag{6.3.16}$$

Its doublet extension is

$$F = X_1(X_1 + X_2) = \{x_1, \ (1,0)^T\}(\{x_1, \ (1,0)^T\} + \{x_2, \ (0,1)^T\}).$$

Using (6.3.1) and (6.3.2), the calculation of F for any value of x is as follows.

$$\begin{aligned} F &= \{x_1, \ (1,0)^T\}(\{x_1 + x_2, \ (1,1)^T\} \\ &= \{x_1(x_1 + x_2), \ (x_1, \ x_1)^T + (x_1 + x_2, \ 0)^T\} \\ &= \{x_1(x_1 + x_2), \ (2x_1 + x_2, \ x_1)^T\}. \end{aligned}$$

Hence the doublet F contains the correct value and gradient vector for the function (6.3.16).

The ideas of forward accumulation are quite easily implemented in programming languages which support overloaded operations for user-defined data types. The code for evaluating the objective function can simply be written in terms of a doublet data type rather than a standard real variable and then, for any values of the independent variables, the numerical value of the first partial derivatives will be returned in the gradient part of the doublet result. This facility is included in OPTIMA which uses forward accumulation for computing first derivatives. OPTIMA also includes procedures for estimating derivatives by finite differences which are used to obtain second derivatives when these are required.

We have only given a brief introduction to an important topic and for more information on automatic differentiation the reader is referred to [26, 32].

Exercises

1. Work through the forward accumulation approach to evaluate the gradient of the function (1.1.5) at $x_2 = 2$.
2. Use the forward accumulation approach to obtain the first partial derivatives of the function in Problem TD1 when $x_1 = x_2 = 1$.

6.4.   Computational costs of derivatives

Finite difference approximations or automatic differentiation software tools allow us to avoid the time and trouble of producing hand-crafted expressions for derivatives and then coding them. However, they may both incur a computational cost. The cost of a forward difference estimate of the gradient of an n-variable function is about the same as n evaluations of the function itself. A central difference estimate of the gradient costs about $2n$ function evaluations. Obtaining a gradient by forward accumulation in doublet arithmetic can also cost about n times as much as one function evaluation. In contrast to these figures, the evaluation of a skilfully coded hand-crafted gradient may be equivalent to much less than n function calculations. In short, therefore, a program which uses hand-crafted derivatives may require less runtime than one which uses derivatives which are approximated or obtained via forward accumulation.

There is a version of automatic differentiation called *reverse accumulation* which is potentially much more efficient than forward accumulation. In terms of arithmetic operations, it is able to compute derivatives as efficiently as the best of hand-crafted expressions. It is, however, a more difficult technique to explain than forward accumulation and its arithmetic efficiency is somewhat offset by the fact that it can be expensive in its memory requirements. We do not discuss it any further here, but details can be found in [32].

Iterative schemes
Each repetition
closes on the vital point:
Pit and Pendulum?

When will it converge?
Each false step, like Poe's raven,
cackles "Nevermore!"

Chapter 7

The Steepest Descent Method

7.1.   Introduction

The *steepest descent* method is the simplest of the gradient methods for optimization in n variables. It can be justified by the following geometrical argument. If we want to minimize a function $F(x)$ and if our current trial point is x_k then we can expect to find better points by moving away from x_k along the direction which causes F to decrease most rapidly. This direction of steepest descent is given by the negative gradient. To use a geographical illustration: suppose we are walking on a hillside in thick fog and wish to get to the bottom of the valley. Even though we cannot see ahead, we can still reach our objective if we make sure each step is taken down the local line of greatest slope.

A formal description of the steepest descent method appears below. Here, and in what follows, subscripts on vectors are used to denote iteration numbers. On occasions when we need to refer to the ith element of a vector x_k we use double-subscript notation x_{k_i}.

Steepest descent with perfect line search

Choose an initial estimate, x_0, for the minimum of $F(x)$.
Repeat for $k = 0, 1, 2, \ldots$
  set $p_k = -\nabla F(x_k)$
  calculate s^* to minimize $\varphi(s) = F(x_k + sp_k)$
  set $x_{k+1} = x_k + s^* p_k$
until $\|\nabla F(x_{k+1})\|$ is sufficiently small.

The one-dimensional minimization in this algorithm can be performed using methods discussed in Chapter 2.

M. Bartholomew-Biggs, *Nonlinear Optimization with Engineering Applications*,
DOI: 10.1007/978-0-387-78723-7_7, © Springer Science+Business Media, LLC 2008

It should be said at once that the steepest descent algorithm is not a particularly efficient minimization method. (The simple strategy of proceeding along the negative gradient works well for functions with near-circular contours; but practical optimization problems may involve functions with narrow curving valleys which need a more sophisticated approach.) However, we consider it at some length because it introduces a pattern common to many optimization methods. In this pattern, an iteration consists of two parts: the choice of a *search direction* (p_k) followed by a *line search* to find a suitable stepsize s^*.

7.2.   Line searches

Definition   A line search which chooses s^* to minimize $\varphi(s) = F(x_k + sp_k)$ is said to be *perfect* or *exact*.

Definition A *weak* or *inexact* line search is one which accepts any value of s such that $F(x_k + sp_k) - F(x_k)$ is negative and bounded away from zero.

A perfect line search gives the greatest possible reduction in F along the search direction. However, as we show later, it may be computationally expensive to do an accurate minimization of $\varphi(s)$ on every iteration. Hence weak searches are often preferred in practice. A convergence proof for the steepest descent algorithm with a weak line search is given later in this chapter.

Line searches play an important part in optimization. If p denotes any search direction and if we write

$$\varphi(s) = F(x_k + sp) \qquad (7.2.1)$$

then, using a Taylor expansion,

$$\varphi(s) = F(x_k) + sp^T \nabla F(x_k) + \frac{s^2}{2} p^T \nabla^2 F(x_k) p + O(s^3 ||p||^3)$$

and so

$$\frac{d\varphi}{ds} = p^T \nabla F(x_k) + sp^T \nabla^2 F(x_k) p + O(s^2 ||p||^3).$$

But

$$\nabla F(x_k + sp) = \nabla F(x_k) + s \nabla^2 F(x_k) p + O(s^2 ||p||^2)$$

and so

$$\frac{d\varphi}{ds} = p^T \nabla F(x_k + sp). \qquad (7.2.2)$$

(We can also derive this relationship by using the chain rule.) From (7.2.2) we deduce that the initial slope, as we move away from x_k along the search direction p, is given by $p^T \nabla F(x_k)$.

Definition The vector p is a *descent direction* with respect to the function $F(x)$ at the point x_k if it satisfies the condition

$$p^T \nabla F(x_k) < 0. \qquad (7.2.3)$$

If (7.2.3) holds then p is a suitable search direction for an iteration of a minimization algorithm which begins at x_k.

Proposition If s^* is the step which minimizes $\varphi(s)$ then

$$p^T \nabla F(x_k + s^* p) = 0. \qquad (7.2.4)$$

Proof The result follows on putting $d\varphi/ds = 0$ on the left of (7.2.2).

Condition (7.2.4) means that a perfect line search terminates at a point where the gradient vector is orthogonal to the direction of search.

A steepest descent example

We now apply steepest descent to the function

$$F(x) = (x_1 - 1)^2 + x_2^3 - x_1 x_2.$$

The gradient is $g = (2x_1 - 2 - x_2, \; 3x_2^2 - x_1)^T$. Hence, if we take $x_0 = (1, 1)^T$ then $F_0 = 0$ and $g_0 = (-1, 2)^T$. On the first iteration, the search direction is $p_0 = -g_0$ and the new solution estimate is

$$x = x_0 + sp_0 = (1, \; 1)^T + s(1, \; -2)^T = (1 + s, \; 1 - 2s)^T.$$

We can use (7.2.3) to confirm that p_0 is a descent direction because

$$p_0^T g_0 = (1, \; -2) \begin{pmatrix} -1 \\ 2 \end{pmatrix} = -5.$$

Now we want to find s to minimize

$$\varphi(s) = F(x_0 + sp_0) = s^2 + (1 - 2s)^3 - (1 + s)(1 - 2s). \qquad (7.2.5)$$

To find a steplength s^* to minimize $\varphi(s)$ we solve $d\varphi/ds = 0$. This leads to

$$2s - 6(1 - 2s)^2 - (1 - 2s) + 2(1 + s) = -24s^2 + 30s - 5 = 0.$$

(In this case we have a quadratic equation which can be solved analytically. In general, however, s^* must be found by an iterative method such as bisection.)

On solving $24s^2 - 30s + 5 = 0$ we find that the smaller root $s^* \approx 0.198$ gives the minimum of φ (and the larger root corresponds to a maximum). Hence a perfect search will give the new point $x = (1.198, \ 0.604)^T$.

A second steepest descent iteration from $x_1 = (1.198, \ 0.604)^T$ will use a search direction $p_1 = -g_1 = (0.208, \ 0.1036)^T$. The new point will then be

$$x_2 = (1.198 + 0.208s, \ 0.604 + 0.1036s)^T$$

where s is again chosen by a perfect line search. Continuing in this way, we can expect that a minimum will be found if enough iterations are performed.

Exercises

1. In the first iteration of the worked example above, show that the same value of s^* would be obtained by solving $p_0^T g = 0$ where g is calculated at $x = (1 + s, \ 1 - 2s)^T$.

2. Perform another iteration of the steepest descent method with perfect line searches applied to (4.1.6) following on from the point $x_1 = (1.198, \ 0.604)^T$.

3. Show that the steepest descent method with perfect line searches generates successive search directions that are orthogonal.

4. Do one iteration of steepest descent for $F(x) = 2x_1^2 + 3x_1 x_2 + 5x_2^2 - x_1$ starting from $x_1 = x_2 = 0$.

7.3.   Convergence of the steepest descent method

Experience shows that methods using perfect line searches may not make much better overall progress than those using weak searches. The following result shows that the steepest descent method with a weak line search can converge to a stationary point.

Proposition   Let $F(x)$ be a function which is twice continuously differentiable and bounded below. Also let its Hessian matrix be bounded, so that for some positive scalar M,

$$z^T \nabla^2 F(x) z \le M||z||^2$$

for any vector z. Then a sequence of steepest descent iterations

$$x_{k+1} = x_k - \frac{1}{M} \nabla F(x_k)$$

(i.e., which use a constant stepsize $s = M^{-1}$) will produce a sequence of points x_k such that $||\nabla F(x_k)|| \to 0$ as $k \to \infty$.

Proof Suppose the statement is false and that, for some positive ϵ,

$$||\nabla F(x_k)|| > \epsilon \quad \text{for all } k.$$

Now consider a typical iteration starting from a point x where $p = -\nabla F(x)$ and $x^+ = x + sp$. By the mean value theorem, for some ξ between x and x^+,

$$F^+ - F = sp^T \nabla F(x) + \frac{1}{2} s^2 p^T \nabla^2 F(\xi) p. \tag{7.3.1}$$

Hence, writing g for $\nabla F(x)$,

$$F^+ - F = -\frac{g^T g}{M} + \frac{g^T \nabla^2 F(\xi) g}{2M^2} \le -\frac{g^T g}{M} + \frac{g^T g}{2M},$$

using the bound on $\nabla^2 F$. Now by the assumption at the start of the proof we have, on every iteration

$$F^+ - F \le -\frac{\epsilon^2}{2M}. \tag{7.3.2}$$

But if this holds for an infinite number of steps it contradicts the fact that $F(x)$ is bounded below; and hence our initial assumption must be false and there exists an integer K such that $||\nabla F(x_k)|| \le \epsilon$ for all $k > K$.

The above proposition does not relate to an algorithm which is either practical or efficient. We would not in general be able to determine the constant M and, even if we could, the stepsize $s = M^{-1}$ would usually be much less than the perfect step and convergence would be slow. However we can use the same *reductio ad absurdum* approach to show the convergence of the steepest descent algorithm as stated in Section 7.1.

Corollary If the function $F(x)$ satisfies the conditions of the preceding proposition then the steepest descent algorithm with perfect line searches produces a sequence of points x_k such that $||\nabla F(x_k)|| \to 0$ as $k \to \infty$.

Proof This result follows because, with a perfect line search, the decrease in function value obtained on every iteration must be at least as good as that given by the bound (7.3.2). Therefore it would still imply a contradiction of F being bounded below if the iterations did not approach a stationary point.

The rate of convergence of steepest descent

The fact that an algorithm can be proved to converge does not necessarily imply that it is a good method. Steepest descent, whether using perfect or weak line searches, is not usually to be recommended in comparison with the algorithms introduced in later chapters. This is because its rate of convergence can be slow, as shown in the next example. Consider the problem

$$\text{Minimize} \quad F(x) = \frac{1}{2}(x_1^2 + qx_2^2).$$

Then $\nabla F = (x_1, qx_2)^T$ and F has a minimum at $x^* = (0, 0)^T$. (A simple sketch shows that the contours are ellipses.)

If we choose $x_0 = (1, q^{-1})^T$ as a starting point then $p_0 = (-1, -1)^T$. Thus the next iterate will be of the form

$$x_1 = (1 - s, \ q^{-1} - s)^T. \tag{7.3.3}$$

A perfect line search finds s so that $p_0^T g_1 = 0$, which implies

$$(-1, -1) \begin{pmatrix} 1 - s \\ 1 - qs \end{pmatrix} = -1 + s - 1 + qs = 0$$

and so $s = 2(1 + q)^{-1}$. Substituting in (7.3.3) we get the new point

$$x_1 = K(1, -q^{-1})^T \quad \text{where} \quad K = \frac{(q - 1)}{(q + 1)}. \tag{7.3.4}$$

It follows from (7.3.4) that $||x_1|| = K||x_0||$ and so the error after the first iteration is K times the error at the starting point. In a similar way we can show

$$||x_2|| = K||x_1|| = K^2||x_0||. \tag{7.3.5}$$

In the special case when $q = 1$ (when F has circular contours) steepest descent performs well, because K is zero and the solution is found in one iteration. For larger values of q, however, (7.3.5) shows that the solution error decreases by a constant factor K on each iteration. Moreover, K is close to 1 for quite moderate values of q. For instance, $K \approx 0.82$ when $q = 10$ and so about 60 iterations would be needed to reduce $||x||$ to around 10^{-5}. Convergence would be yet slower for $q = 100$. This example illustrates a general property of the steepest descent method, which we state without proof.

Proposition If $F(x)$ is a function for which the steepest descent algorithm converges to a stationary point x^*, then there exists an integer $\bar{k}$

and a positive real constant $K(< 1)$ such that, for $k > \bar{k}$,

$$||x_{k+1} - x^*|| < K||x_k - x^*||.$$

This means that the steepest descent method generally displays *linear* convergence near the solution, with the errors in the approximate minima, x_k, decreasing by a constant factor on every iteration.

7.4.   Results with steepest descent

Performance of the steepest descent method can be illustrated by results for Problems TD1–OC2. The OPTIMA implementation of the steepest descent method with perfect line searches, is denoted by SDp and Table 7.1 gives the numbers of iterations and function evaluations needed to satisfy convergence test (4.3.2) when $\epsilon = 10^{-4}, 10^{-5}$ and 10^{-6}. Table 7.1

Problem	Low Accuracy itns/fns	Standard Accuracy itns/fns	High Accuracy itns/fns
TD1	18/71	24/83	28/91
TD2	16/48	20/56	24/64
VLS1	4/9	5/11	5/11
TLS1	26/69	38/93	48/113
VLS2	20/108	25/118	29/126
R1(1)	43/294	43/294	44/296
R1(2)	1608/7650	2977/12673	3555/14560
OC1(4)	417/835	609/1219	801/1603
OC2(4)	111/241	153/325	197/413

Table 7.1.   SDp solutions for Problems TD1–OC2.

shows that SDp performs in a similar way on Problems TD1 and TD2, taking four iterations to reduce the gradient norm by an order of magnitude, from $O(10^{-5})$ to $O(10^{-6})$. Hence its ultimate rate of convergence is given approximately by $||x_{k+1}-x^*|| < 0.56||x_k-x^*||$ because $0.56^4 \approx 0.1$.

Among the least-squares problems, SDp clearly finds TLS1 the most difficult. Problem VLS2 is also quite challenging but VLS1 with its simple quadratic objective function proves relatively easy. On Problem TLS1, SDp takes about 12 iterations to reduce the gradient by an order of magnitude and so the ultimate convergence rate is approximately $||x_{k+1} - x^*|| < 0.83||x_k - x^*||$. The most extreme examples of slow convergence near a solution are seen on Problems OC1(4) and R1(2).

It is interesting to compare the performance of steepest descent with that of the direct search methods shown in Table 5.1. If we take numbers of function evaluations as a basis for comparison we see that, in all cases except R1(2), SDp outperforms UNIVAR. It also does better than DIRECT on all problems except R1(1) and R1(2). In terms of function evaluations, however, SDp only beats the simplex method on two of the problems and does appreciably worse on the routing and optimal control examples.

Exercises

1. Apply SDp to Problem TD1 with the starting guess $x_1 = x_2 = 1$ and explain what happens.
2. Apply SDp to a modified form of Problem TD2 in which the specified surface area is $S^* = 30$.
3. Apply SDp to a modified form of Problem VLS2 in which there is an extra data point $(5, 0.1)$.
4. Perform numerical experiments applying SDp to Problem R1(2) to see how close the initial guess must be to the solution in order for the method to converge in fewer than 1000 iterations.
5. Apply SDp to the problems R1(1) and R1(2) with $\rho = 0.01$ and comment on the solutions.
6. Investigate the performance of SDp on Problems OC1(6) and OC2(6).

Chapter 8

Weak Line Searches and Convergence

8.1. Wolfe's convergence theorem

The steepest descent method introduces some important ideas which are common to many other minimization techniques. These are (i) the choice of a search direction, p, to satisfy the descent property (7.2.3); and (ii) the use of a line search to ensure that the step, s, along p decreases the function. Optimization techniques differ mainly in the way that p is calculated on each iteration.

Wolfe's theorem [66] gives precise conditions on p and s which guarantee convergence of any minimization algorithm. We now define these Wolfe conditions with x_k denoting an estimate of the minimum of $F(x)$ and $g_k = \nabla F(x_k)$.

Definition The *first Wolfe condition* is a stronger form of (7.2.3), namely

$$p^T g_k \leq -\eta_0 ||p|| \, ||g_k||, \qquad (8.1.1)$$

where η_0 is a small positive constant, typically $\eta_0 = 0.01$.

If (8.1.1) holds then θ, the angle between p and $-g_k$, is such that $\cos \theta$ is positive and bounded away from zero. In other words $-\pi/2 < \theta < \pi/2$.

Before stating the other Wolfe conditions, we let $\bar{s}$ denote the steplength along p for which

$$\varphi(\bar{s}) = \varphi(0) \quad \text{which is eqivalent to} \quad F(x_k + \bar{s}p) = F(x_k). \qquad (8.1.2)$$

Clearly $\bar{s} > s^*$, the step which minimizes φ; and on a quadratic function we can show that $\bar{s} = 2s^*$ (see Exercise 3, below). The purpose of the next two Wolfe conditions is to define an acceptable step s as being one which is neither too long (i.e., too close to $\bar{s}$) or too short (too near zero).

M. Bartholomew-Biggs, *Nonlinear Optimization with Engineering Applications*,
DOI: 10.1007/978-0-387-78723-7_8, © Springer Science+Business Media, LLC 2008

Definition The *second Wolfe condition* is

$$F(x_k + sp) - F(x_k) \leq \eta_1 \ s \ p^T g_k, \tag{8.1.3}$$

for some constant η_1, such that $0.5 > \eta_1 > 0$ (typically $\eta_1 = 0.1$).

Condition (8.1.3) ensures that the step taken produces a nontrivial reduction in the objective function and hence that s is bounded away from $\bar{s}$.

Definition The *third Wolfe condition* is

$$|F(x_k + sp) - F(x_k) - s \ p^T g_k| \geq \eta_2 |s \ p^T g_k|, \tag{8.1.4}$$

where η_2 is a constant such that $0.5 > \eta_2 > 0$.

The inequality (8.1.4) ensures s is bounded away from zero by requiring the actual decrease in F to be bounded away from the linear predicted reduction.

Wolfe's Theorem [66] If $F(x)$ is bounded below and has bounded second derivatives then any minimization algorithm which satisfies (8.1.1)–(8.1.4) on a regular subsequence of iterations (and does not allow F to increase) will terminate in a finite number of iterations at a point where $||\nabla F(x)||$ is less than any chosen positive tolerance.

A proof of Wolfe's theorem is not given here but it is similar to that for the steepest descent convergence result in Section 7.3. To explain the second and third Wolfe conditions we introduce the ratio

$$D(s) = \frac{F(x_k + sp) - F(x_k)}{s \ p^T g_k}. \tag{8.1.5}$$

Figure 8.1 illustrates the behaviour of this function when F is quadratic.

Clearly $D(\bar{s}) = 0$ and we can also show that $D(s) \to 1$ as $s \to 0$. Moreover, if $F(x)$ is a quadratic function, $D(s)$ decreases linearly from 1 to 0 as s increases from 0 to $\bar{s}$. In particular $D(s^*) = 0.5$. The second Wolfe condition is equivalent to the requirement that $D(s) \geq \eta_1$. Similarly, the third Wolfe condition holds if $D(s) \leq 1 - \eta_2$. The vertical dashed lines in Figure 8.1 indicate an acceptable range for s when $\eta_1 = \eta_2 = 0.1$.

If $F(x)$ is nonquadratic then $D(s)$ will not be linear. However, if F is convex, D will still lie in the range $1 \geq D(s) \geq 0$ when $0 \leq s \leq \bar{s}$. Figure 8.2 illustrates the acceptable range for s on a cubic function when $\eta_1 = \eta_2 = 0.1$. Note that the left- and right-hand excluded regions are not now the same size.

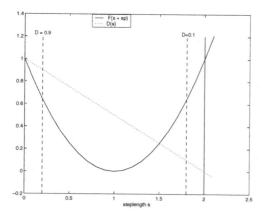

Figure 8.1. Wolfe conditions on a quadratic function.

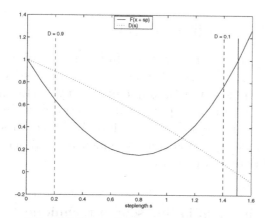

Figure 8.2. Wolfe conditions on a nonquadratic function.

If $F(x)$ is nonconvex then the ratio $D(s)$ may exceed 1 in the range $0 \leq s \leq \bar{s}$. This can be seen in Figure 8.3, where the function has slight negative curvature near to $s = 0$. The third Wolfe condition – that the step s must not be too close to zero – can then be expressed as $|1 - D(s)| \geq \eta_2$. Figure 8.3 shows the acceptable range for s when $\eta_1 = \eta_2 = 0.1$.

Exercises

1. Prove that the steepest descent direction satisfies Wolfe condition 1.
2. Use Taylor series to show that $D(s) \rightarrow 1$ as $s \rightarrow 0$.
3. Use (8.1.2) to prove that $\bar{s} = 2s^*$ when F is quadratic.
4. An alternative to (8.1.4) is $|p^T g(x_k + sp)| \leq \eta_3 |p^T g(x_k)|$ for some constant $\eta_3 (1 > \eta_3 > 0)$. Explain why this causes s to be bounded away from zero.
5. If $F(x)$ is quadratic, prove that (8.1.5) implies $D(s^*) = 0.5$.

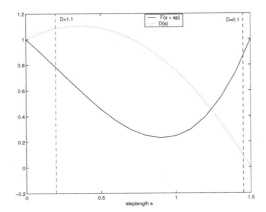

Figure 8.3.   Wolfe conditions on a nonconvex nonquadratic function.

8.2.   The Armijo line search

Conditions (8.1.3) and (8.1.4) will be fulfilled if s minimises the line search function $\varphi(s)$ in (7.2.1). However, they also justify the use of a weak line search. This could be implemented by performing an exact search with a low-accuracy stopping rule. However, a simpler form of weak search, based on the second and third Wolfe conditions, is called the *Armijo technique* [1]. This can conveniently be described in terms of the ratio $D(s)$ defined by (8.1.5).

The Armijo line search technique

Let p be a search direction satisfying (8.1.1)
Choose constants $C > 1$, $c < 1$ and η_1, η_2 such that $0 < \eta_1, \eta_2 < 0.5$
Set $s = 1$ and $s_{\min} = 0$
Repeat for $j = 0, 1, 2, \ldots$
If $|1 - D(s)| < \eta_2$ then set $s_{\min} = s$ and $s = Cs$
else exit
Repeat for $k = 0, 1, 2, \ldots$
set $s = s_{\min} + c(s - s_{\min})$
until $D(s) \geq \eta_1$

This search first ensures that s is "big enough" and then modifies s if necessary to produce a decrease in F consistent with (8.1.3). The second, step-reducing, phase must not cause a violation of (8.1.4).

The Armijo search can be modified to perform exact minimizations when F is quadratic by making use of the fact that $D(s^*) = 0.5$. This modified search uses linear interpolation or extrapolation to estimate a value of s such that $D(s) = 0.5$ (but using the constants C and c to prevent excessively large or small corrections).

A modified Armijo line search

Let p be a search direction satisfying (8.1.1)
Choose constants $C > 1$, $c < 1$ and η_1, η_2 such that $0 < \eta_1, \eta_2 < 0.5$
Set $s = 1$ and $s_{\min} = 0$.
Repeat for $j = 0, 1, 2, \ldots$
if $|1 - D(s)| < \eta 2$ then set

$$s_{\min} = s \quad \text{and} \quad s = \min\left(Cs, \frac{0.5s}{1 - D(s)}\right)$$

else exit
Repeat for $k = 0, 1, 2, \ldots$

$$s = \max\left(s_{\min} + c(s - s_{\min}), \frac{0.5s}{1 - D(s)}\right)$$

until $D(s) \geq \eta_1$

For nonquadratic functions, the Armijo search with $\eta_1, \eta_2 \approx 0.5$ will (usually) give a better estimate of the line minimum than one which uses $\eta_1, \eta_2 \approx 0.1$. However, in order to perform a perfect line search on nonquadratic functions the OPTIMA software uses an Armijo search to obtain an approximation to the line minimum, and then switches to the secant method in order to locate a point where the directional derivative $\varphi'(s) = p^T g$ is close enough to zero.

Exercises

1. Calculate the point found on the first iteration of the example in Section 7.2 if the line search is done by the Armijo method with $\eta_1 = \eta_2 = 0.1$, using $s = 1$ as the first trial step. What point would be found if the modified Armijo search were used instead?
2. Write and test a procedure implementing the simple Armijo search.

8.3. Further results with steepest descent

We now consider the performance of SDw, the OPTIMA implementation of steepest descent with a weak line search. The entries in Table 8.1 are numbers of iterations and function calls needed to satisfy the stopping rule (4.3.2) with $\epsilon = 10^{-4}, 10^{-5}$ and 10^{-6}.

Problem	Low Accuracy itns/fns	Standard Accuracy itns/fns	High Accuracy itns/fns
TD1	22/46	28/58	34/70
TD2	14/27	20/39	26/51
VLS1	4/9	5/11	5/11
TLS1	50/92	61/105	69/114
VLS2	17/52	21/60	25/68
R1(1)	90/291	104/326	118/361
R1(2)	282/824	1446/3734	2620/6669
OC1(4)	417/835	609/1219	801/1603
OC2(4)	119/239	165/331	211/423

Table 8.1.   SDw solutions for Problems TD1–OC2.

By comparing Tables 7.1 and 8.1 we see that a weak line search is usually preferable to a perfect one. For Problems VLS1 and OC1, both methods give the same performance; but SDw uses fewer function calls (although more iterations) than SDp when applied to Problem TD1. On Problem TD2, SDw and SDp use similar numbers of iterations but the weak search makes fewer function calls. On Problem TLS1, SDw takes many more iterations than SDp but still uses fewer function calls overall (about 1.5 per iteration as opposed to between 2 and 3). It is only on Problems R1(1) and OC2 that SDw is inferior to SDp in terms of both iterations and function calls.

In comparison with the numbers of function evaluations needed by the direct search methods (Table 5.1), SDw is more competitive than SDp and outperforms the simplex method on the first five problems.

Exercises
1. Change the starting point for TLS1 to $x_1 = x_2 = 1$ and compare the performance of SDp and SDw.
2. Apply SDw to a modified form of Problem TD2 in which the specified surface area is $S^* = 30$.
3. Apply SDw to a modified form of Problem VLS2 in which there is an extra data point (5, 0.1).
4. Use the results in Table 8.1 to estimate the rate of convergence of SDw on each problem.
5. Perform numerical experiments applying SDw to Problem R1(2) to see how close the initial guess must be to the solution in order for the method to converge in fewer than 1000 iterations.
6. Apply SDw to the problems R1(1) and R1(2) with $\rho = 0.01$ and comment on the solutions.
7. Apply SDw to Problems OC1(6) and OC1(8).
8. Write and test a procedure implementing the steepest descent method using a weak, Armijo-type line search.

9. Consider a version of the steepest descent method in which the iteration has the form $x_{k+1} = x_k - hg_k$ where h is a small positive constant on every step. The points traced out by such an iteration are an approximation to the continuous steepest descent path obtained by solving the ordinary differential equation

$$\frac{dx}{dt} = -\nabla F(x)$$

with the initial condition $x = x_0$ when $t = 0$. By modifying an OPTIMA procedure (or otherwise) implement this algorithm and investigate its performance on some of the test examples TD1–OC2.

Calculated risk [2]

It was 6 a.m. at Heathrow when
their machine was woken up by mine.
Turn it on, the stern attendant said
and prove to us it's what it seems to be:
a harmless, necessary calculator.

Feeling rather smug to be accused
of something I was sure I had not done,
I rattled buttons to evaluate
suspicion and obtained precisely nothing –
a single open zero with no countdown.

Permitted to proceed beyond the gate,
I felt my reservations multiply
as fast as bits of ticket were subtracted.
A formula that proves one's innocence
in terms of integers may miss the point.

Chapter 9

Newton and Newton-like Methods

9.1. Quadratic models and the Newton step

The steepest descent algorithm performs badly on the example in
Section 7.3 chiefly because it uses no second derivative information.
More effective methods are based on the properties of a quadratic func-
tion,

$$Q(x) = \frac{1}{2}x^T A x + b^T x + c \qquad (9.1.1)$$

where A is a constant $n \times n$ matrix, b a constant n-vector and c a constant
scalar. The gradient and Hessian of Q are

$$\nabla Q(x) = Ax + b \quad \text{and} \quad \nabla^2 Q(x) = A. \qquad (9.1.2)$$

If A is nonsingular then (9.1.1) has a unique stationary point which is
found by solving the linear equations

$$Ax = -b. \qquad (9.1.3)$$

The solution will be a minimum if the Hessian matrix A is positive-
definite. On the other hand, it will be a maximum if A is negative-
definite or a saddle point if A is indefinite.

We can also apply these ideas to a nonquadratic function $F(x)$.
Suppose x_k is an estimate of the minimum of $F(x)$ and that $g_k = \nabla F(x_k)$, $G_k = \nabla^2 F(x_k)$. We can approximate F and ∇F by the
truncated Taylor series

$$F(x_k + p) \approx Q(p) = F(x_k) + p^T g_k + \frac{1}{2}(p^T G_k p) \qquad (9.1.4)$$

and

$$\nabla F(x_k + p) \approx \nabla Q(p) = g_k + G_k p. \qquad (9.1.5)$$

M. Bartholomew-Biggs, *Nonlinear Optimization with Engineering Applications*,
DOI: 10.1007/978-0-387-78723-7_9, © Springer Science+Business Media, LLC 2008

Therefore, if G_k is positive-definite, a first-order estimate of the step from x_k to the minimum of F is

$$p = -G_k^{-1} g_k. \qquad (9.1.6)$$

This leads to the following algorithm.

The Newton method

Choose x_0 as an initial estimate of the minimum of $F(x)$
Repeat for $k = 0, 1, 2, \ldots$
Set $g_k = \nabla F(x_k)$, $G_k = \nabla^2 F(x_k)$.
if G_k is positive-definite then obtain p_k by solving $G_k p_k = -g_k$
else set $p_k = -g_k$
Find s so $F(x_k + s p_k)$ satisfies (8.1.3), (8.1.4) for some η_1, η_2
Set $x_{k+1} = x_k + s p_k$
until $\|\nabla F(x_{k+1})\|$ is sufficiently small.

The vector p_k given by (9.1.6) is called the *Newton correction* and is based on regarding Q as a local quadratic model of F. Under favourable conditions – that is, when the Hessian of F is positive-definite – the Newton algorithm can be very efficient. The "natural" steplength implied by the quadratic model is $s = 1$; and in practice this often satisfies the Wolfe conditions and effectively eliminates the line search.

If G_k is not positive-definite then the quadratic model has a maximum or a saddle point rather than a minimum and then the Newton correction may not be suitable. In this case we need an alternative search direction; and in the algorithm above we simply revert to steepest descent. This is not a particularly good option, however, and we discuss this issue in greater detail in a subsequent section.

A worked example

We can demonstrate a typical Newton iteration using the function

$$F(x) = x_1^4 + 2x_2^2 + x_1 - x_2.$$

We avoid the use of subscripts to denote iteration numbers and so $x = (-\frac{1}{2}, \frac{1}{2})^T$ is the starting point. We refer to the search direction as p and a line search along p will yield $x^+ = x + sp$. Because

$$g(x) = \begin{pmatrix} 4x_1^3 + 1 \\ 4x_2 - 1 \end{pmatrix} = \begin{pmatrix} \frac{1}{2} \\ 1 \end{pmatrix} \quad \text{and}$$

$$G(x) = \begin{pmatrix} 12x_1^2 & 0 \\ 0 & 4 \end{pmatrix} = \begin{pmatrix} 3 & 0 \\ 0 & 4 \end{pmatrix}$$

the Newton correction p is obtained by solving $Gp = -g$, which is

$$3p_1 = -\frac{1}{2} \tag{9.1.7}$$

$$4p_2 = -1. \tag{9.1.8}$$

This gives $p_1 = -\frac{1}{6}$ and $p_2 = -\frac{1}{4}$ and so the new point is of the form

$$x^+ = \left(-\frac{1}{2} - \frac{1}{6}s, \ \frac{1}{2} - \frac{1}{4}s\right)^T.$$

Using the "natural" steplength $s = 1$ we get $x^+ = (-\frac{4}{6}, \frac{1}{4})^T$ and then

$$F(x^+) = \frac{256}{1296} + \frac{2}{16} - \frac{4}{6} - \frac{1}{4} \approx -0.5942.$$

But the value of F at the initial point is -0.4375, and so the step $s = 1$ has produced an acceptable reduction in F.

Exercises

1. In the worked example above, calculate the optimal step s^* for a perfect line search along the Newton direction. What are the largest values of η_1 and η_2 for which the step $s = 1$ satisfies the second and third Wolfe conditions?

2. Do one iteration of the Newton method applied to the function

$$F(x) = (x_1 - 1)^2 + x_2^3 - x_1 x_2 \qquad ,$$

starting from $x_1 = x_2 = 1$. What happens when you start from $x = (1, -1)$?

Positive-definiteness and Cholesky factors

In the Newton algorithm we must determine whether the Hessian G_k is positive-definite. Very conveniently, the test for positive-definiteness can be combined with the solution of $G_k p_k = -g_k$ if we use the method of *Cholesky factorization*. This seeks triangular factors of G_k so that

$$G_k = LL^T \tag{9.1.9}$$

where L is a lower triangular matrix. Once we have found these factors we can solve $G_k p_k = -g_k$ by first obtaining an intermediate vector z to satisfy

$$Lz = -g_k \tag{9.1.10}$$

and then getting p_k from

$$L^T p_k = z. \tag{9.1.11}$$

The two linear systems (9.1.10) and (9.1.11) are easy to solve because they involve triangular coefficient matrices and so z and p_k are obtained by simple forward and backward substitution.

The Cholesky factorization (9.1.9) always exists if G_k is positive definite. Conversely, if G_k is not positive definite then the factorization process will break down. Attempting to find the Cholesky factors is usually the most efficient way of testing a symmetric matrix for positive definiteness.

We now describe the Cholesky method for solving a symmetric positive-definite linear system $Ax = b$ using the factorization $A = LL^T$. The method of calculating L is based on the fact that its elements must satisfy

$$
A = \begin{pmatrix}
l_{11} & 0 & \cdots & 0 & \cdots & 0 \\
l_{21} & l_{22} & \cdots & 0 & \cdots & 0 \\
& & \cdots & \cdots & & \\
l_{k1} & l_{k2} & \cdots & l_{kk} & \cdots & 0 \\
& & \cdots & \cdots & & \\
l_{n1} & l_{n2} & \cdots & l_{nk} & \cdots & l_{nn}
\end{pmatrix}
\begin{pmatrix}
l_{11} & l_{21} & \cdots & l_{k1} & \cdots & l_{n1} \\
0 & l_{22} & \cdots & l_{k2} & \cdots & l_{n2} \\
& & \cdots & \cdots & & \\
0 & 0 & \cdots & l_{kk} & \cdots & l_{nk} \\
& & \cdots & \cdots & & \\
0 & 0 & \cdots & 0 & \cdots & l_{nn}
\end{pmatrix}.
$$

Considering the first row of A, the rules of matrix multiplication imply

$$
a_{11} = l_{11}^2; \quad a_{1j} = l_{11}l_{j1} \quad \text{for } j = 2, \ldots, n.
$$

Hence the first column of L can be obtained from

$$
l_{11} = \sqrt{a_{11}}; \quad l_{j1} = \frac{a_{1j}}{l_{11}} \quad \text{for } j = 2, \ldots, n.
$$

In the second row of A we have

$$
a_{22} = l_{21}^2 + l_{22}^2; \quad a_{2j} = l_{21}l_{j1} + l_{22}l_{j2} \quad \text{for } j = 3, \ldots, n
$$

and so the second column of L is given by

$$
l_{22} = \sqrt{a_{22} - l_{21}^2}; \quad l_{j2} = \frac{(a_{j2} - l_{21}l_{j1})}{l_{22}} \quad \text{for } j = 3, \ldots, n.
$$

More generally, by considering the k-th row of A, we obtain the following expressions for the k-th column of L.

$$
l_{kk} = \sqrt{a_{kk} - \sum_{i=1}^{k-1} l_{ki}^2}; \quad l_{jk} = \frac{(a_{kj} - \sum_{i=1}^{k-1} l_{ki}l_{ji})}{l_{kk}} \quad \text{for } j = k+1, \ldots, n.
$$

$$
(9.1.12)
$$

A complete Cholesky factorization consists of applying (9.1.12) for $k = 1, \ldots, n$. The process breaks down at stage k if the calculation of

l_{kk} involves the square root of a negative number. This will not happen if A is positive-definite.

As an example, consider the equations

$$A = \begin{pmatrix} 10 & 1 & 1 \\ 1 & 8 & 2 \\ 1 & 2 & 20 \end{pmatrix} \begin{pmatrix} x_1 \\ x_2 \\ x_3 \end{pmatrix} = \begin{pmatrix} 10 \\ 7 \\ -17 \end{pmatrix}. \tag{9.1.13}$$

The factorization process (9.1.12) gives

$$l_{11} = \sqrt{10} \approx 3.162; \quad l_{21} = l_{31} \approx \frac{1.0}{3.162} \approx 0.3163;$$

$$l_{22} \approx \sqrt{8.0 - 0.3163^2} \approx 2.811; \quad l_{32} \approx \frac{(2.0 - 0.3163 \times 0.3163)}{2.811} \approx 0.6759;$$

$$l_{33} \approx \sqrt{20.0 - 0.3163^2 - 0.6759^2} \approx \sqrt{19.44} \approx 4.409.$$

(The product LL^T will be found to differ slightly from A because the elements of L have been rounded to four-digit accuracy.)

To solve (9.1.13) we deal first with the lower triangular system $Lz = b$. Forward substitution in

$$\begin{pmatrix} 3.162 & 0 & 0 \\ 0.3163 & 2.811 & 0 \\ 0.3163 & 0.6759 & 4.409 \end{pmatrix} \begin{pmatrix} z_1 \\ z_2 \\ z_3 \end{pmatrix} = \begin{pmatrix} 10 \\ 7 \\ -17 \end{pmatrix}$$

gives

$$z_1 \approx \frac{10}{3.162} \approx 3.163, \quad z_2 \approx \frac{(7 - 0.3163 z_1)}{2.811} \approx 2.134$$

$$\text{and} \quad z_3 \approx \frac{(-17 - 0.6759 z_2 - 0.3163 z_1)}{4.409} \approx 4.409.$$

The upper triangular system $L^T x = z$ is

$$\begin{pmatrix} 3.162 & 0.3163 & 0.3163 \\ 0 & 2.811 & 0.6759 \\ 0 & 0 & 4.409 \end{pmatrix} \begin{pmatrix} x_1 \\ x_2 \\ x_3 \end{pmatrix} = \begin{pmatrix} 3.163 \\ 2.134 \\ 4.409 \end{pmatrix}$$

and backward substitution yields $x \approx (1.0, 1.0, -1.0)^T$. Clearly this satisfies the original system (9.1.13).

Exercises

1. Show that solving (9.1.10) and (9.1.11) yields the solution to $G_k p_k = -g_k$.

2. Solve the system of equations

$$
\begin{array}{rcrcrcl}
10x_1 & + & x_2 & + & x_3 & = & 9 \\
x_1 & + & 8x_2 & + & 2x_3 & = & 11 \\
x_1 & + & 2x_2 & + & 12x_3 & = & -31
\end{array}
\qquad (9.1.14)
$$

using the Cholesky method.

9.2. Newton method advantages and drawbacks

If the eigenvalues of the Hessian matrix $\nabla^2 F$ are bounded away from zero then it can be shown that the Newton iteration produces search directions which satisfy the first Wolfe condition (8.1.1). Therefore the Newton method converges if it uses a line search to satisfy the second and third Wolfe conditions. The rate of convergence can be quadratic, as stated in the following result.

Proposition If $F(x)$ is a function for which the Newton algorithm converges to a local minimum x^* and if the smallest eigenvalue of $\nabla^2 F(x^*)$ is $m > 0$ and if the third derivatives of $F(x)$ are bounded in some neighbourhood of x^* then there exists an integer $\bar{k}$ and a positive real constant $K(< 1)$ such that, for $k > \bar{k}$,

$$
\|x_{k+1} - x^*\| < K\|x_k - x^*\|^2.
$$

(The essentials of the proof are similar to the one-variable case in Chapter 2.)

It is important to point out that such theoretical convergence rates are not always observed in practice because of rounding errors in computer arithmetic. Because all calculated results must be expressed in some fixed number of digits (about 14 in the double precision arithmetic used in OPTIMA) there will inevitably be small errors in computed values of F, ∇F and $\nabla^2 F$ during the solution of an optimization problem. Such errors may become significant when ∇F is near zero and they can prevent iterative methods from reaching solutions with arbitrarily high accuracy. (For a fuller account of rounding errors see the text by Higham [36].)

In spite of the above cautionary remarks, however, the theoretical quadratic convergence of the Newton method does imply that it can be very efficient. Unfortunately, however, the method also has some drawbacks.

(i) Hand crafting of all the required second derivatives can be laborious. Along with the subsequent coding of the derivative expressions it is both time consuming and liable to error. As explained earlier,

this effort can be avoided by the use of finite difference approximations or by extension of the automatic differentiation techniques described in Chapter 6. However, the Newton method has sometimes been neglected because it is perceived as requiring too much effort on the part of a user. *(ii)* The Newton method is computationally expensive because it solves a system of linear equations to obtain the search direction. The Cholesky method is more efficient than the general-purpose Gaussian elimination method but it still uses $O(\frac{1}{6}n^3)$ multiplications per iteration. (A possible way of reducing this would be to perform a low-accuracy solution of $G_k p = -g_k$. This idea leads to the *truncated Newton* method explained in section 11.5.)

(iii) The most serious difficulty for the Newton approach is that it does not invariably provide a good search direction. As we have already mentioned, the Cholesky solution of $G_k p_k = -g_k$ may break down because G_k is not positive-definite. If we were to solve $G_k p_k = -g_k$ by some other method when this happens then the search direction might lead towards a local maximum or saddle point. (We have already noted similar behaviour of the one-variable Newton method in Section 2.4.) The algorithm given in Section 9.1 deals with the possibility of unsuitable search directions by resorting to the steepest descent direction on certain iterations. Better strategies than this can be devised; but the fact still remains that the Newton algorithm, in practice, requires a "fall-back option" to ensure convergence. Further discussion of this appears in the next section.

Exercise

If a search direction is obtained by solving an $n \times n$ linear system $Bp = -g$ and if B is positive-definite show that p is a descent direction. If the eigenvalues of B are bounded above by M and below by m show that Wolfe condition 1 is satisfied with $\eta_0 = \sqrt{(m/M)}$.

9.3.   Search directions from indefinite Hessians

Matrix modification techniques

Suppose that, during a Newton iteration, the Cholesky factorization breaks down at step k because the calculation of l_{kk} in (9.1.12) involves the square root of a negative argument. We could try to continue with the calculation of a search direction by modifying the Hessian and increasing its k-th diagonal term g_{kk} so that it exceeds $\sum_{i=1}^{k-1} l_{ki}^2$. We would then obtain factors L, L^T of a matrix $\hat{G}_k$ which differs from G_k in one (or more) diagonal elements. These triangular factors can then be used to solve $\hat{G}_k p = -g_k$ and the resulting p will be a descent direction, based on partially correct second derivative information, which could be used instead of the Newton direction.

An alternative strategy when the Cholesky factorization fails is based on the Gershgorin disk theorem [38] which implies that a symmetric matrix A will have all positive eigenvalues if it satisfies

$$a_{kk} > \sum_{j=1,\ j\neq k}^{n} |a_{kj}|$$

(which is called a diagonal dominance condition). Hence, if the Cholesky factorization breaks down we can obtain a modified Hessian $\hat{G}$ by increasing the diagonal elements of G, where necessary, so that

$$g_{kk} \geq M \sum_{j=1,\ j\neq k}^{n} |g_{kj}|$$

for some value of $M > 1$.

In practice, the modifications of diagonal elements of G_k outlined in the previous paragraphs do not always work very well because the computed L and L^T factors may contain very large elements. Gill and Murray [30] and Schnabel and Eskow [60] have suggested more complicated – but more numerically stable – ways of changing the Hessian during the Cholesky process so as to get L and L^T as factors of a matrix $\hat{G}_k$ which may differ from G_k in both diagonal and off-diagonal terms.

Trust-region methods

If G_k is not positive-definite we can get a downhill search direction by solving

$$(\lambda I + G_k)p_k = -g_k \qquad (9.3.1)$$

for a "suitably large" value of the positive scalar λ. This is because the eigenvalues of $(\lambda I + G_k)$ exceed those of G_k by λ and hence, when λ is big enough, $(\lambda I + G_k)$ must be positive-definite.

Using a search direction given by (9.3.1) might seem as arbitrary as the matrix modification approaches outlined above. However, it turns out that (9.3.1) gives p_k as a solution of a subproblem of the form

$$\text{Minimize} \quad Q(p) = \frac{1}{2}p^T G_k p + p^T g_k \quad \text{subject to} \quad ||p||_2 \leq \Delta. \qquad (9.3.2)$$

In other words, $x_k + p_k$ minimizes a quadratic approximation to F subject to an upper bound on the size of the move away from x_k. (The connection between (9.3.2) and (9.3.1) is established in a later chapter.)

Problem (9.3.2) always has a solution even when G_k is not positive-definite. This is because it simply amounts to finding the smallest value of $Q(p)$ within a hyperspherical region around x_k. Hence (9.3.2) provides a reasonable way of choosing a search direction when the Newton correction is not appropriate.

Problem (9.3.2) is the basis for a class of minimization techniques known as *trust-region methods* fully described by Conn *et al.* [14]. Most of the methods that we consider work by choosing a promising search direction first and then determining a stepsize by a line search. Trust-region methods reverse this approach and decide on a suitable stepsize before calculating a direction in which to take it.

Suppose for instance that we have some reason to trust that a local quadratic model of F will be reasonably accurate within a distance $||\Delta||$ of the current iterate x_k. A new point $x_{k+1} = x_k + p_k$ could then be obtained by solving (9.3.2) whether G_k is positive-definite or not. The *trust-region radius*, Δ, can be adjusted from iteration to iteration. It is increased if the actual change $F(x_{k+1}) - F(x_k)$ agrees well with the predicted change in the quadratic model. Conversely, it is decreased if the actual and predicted changes are too inconsistent. Ultimately, the value of Δ becomes large enough for subproblem (9.3.2) to allow full Newton steps to be taken and hence permit quadratic convergence.

One disadvantage of the trust-region approach is that (9.3.2) can be difficult and expensive to solve accurately on each iteration. The relationship between the trust region radius Δ and the value of λ in (9.3.1) is highly nonlinear and it is not easy to obtain p_k to solve (9.3.2) via a single solution of (9.3.1). Therefore most implementations make do with an approximate solution.

We can seek an approximate solution of (9.3.1) in the following way. If we define $\mu = \lambda^{-1}$ then (9.3.1) is equivalent to

$$(I + \mu G)p = -\mu g. \tag{9.3.3}$$

If λ is sufficiently large that the matrix μG is small compared with I, then we can use the expansion

$$(I + \mu G)^{-1} = I - \mu G + \mu^2 G^2 - \mu^3 G^3 + O(\mu^4)$$

to obtain an approximate solution of (9.3.3) as

$$p = -\mu\gamma_0 + \mu^2\gamma_1 - \mu^3\gamma_2 \tag{9.3.4}$$

where

$$\gamma_0 = g, \quad \gamma_k = G^k g, \quad k = 1, 2, \ldots.$$

A suitable value for μ can be found by a trial-and-error process. For a sequence of values $\mu_1, \ldots, \mu_k$ with $0 < \mu_1 < \mu_2 \cdots < \mu_k < ||G||^{-1}$ we use (9.3.4) to obtain $p_1, \ldots, p_k$ as candidate directions (without the cost of any matrix factorization). We then evaluate the curvature of the objective function along each p_i as

$$\kappa_i = \frac{p_i^T G p_i}{p_i^T p_i}.$$

If possible, we then pick a search direction p_i giving the most negative value κ_i. Otherwise we choose the one with the smallest positive value. In either case we expect quite a large step to be taken which may cause the search to move rapidly away from the nonconvex region to one where G is positive-definite.

Another way of avoiding the cost of solving (9.3.2) is to reduce it to a 2-D problem. For instance we could combine the negative gradient $-g_k$ with the Newton direction p_k (even if this is uphill) [6]. This would mean seeking a new point $x_{k+1} = x_k - \alpha g_k + \beta p_k$ that gives the least value of F in the plane $(p_k, -g_k)$, subject to a restriction on stepsize. Better still, for the non positive-definite case, would be to determine a direction of *negative curvature*, v, such that $v^T G_k v < 0$. We could then search for the minimum of F in a plane defined by $-g_k$ and v (see [68]).

Exercises

1. If $\bar{G} = \lambda I + G_k$ show that $x^T \bar{G} x > 0$ for all $x \neq 0$, when λ is sufficiently large.

2. Show that the matrix

$$G = \begin{pmatrix} 2 & 0.1 \\ 0.1 & -1 \end{pmatrix}$$

is not positive-definite and then test the accuracy of (9.3.4) as an approximate solution of

$$\begin{pmatrix} 1 + 2\mu & 0.1 \\ 0.1 & 1 - \mu \end{pmatrix} \begin{pmatrix} p_1 \\ p_2 \end{pmatrix} = \begin{pmatrix} -1 \\ -2 \end{pmatrix}$$

when $\mu = 1.1, 1.3, 1.5$.

3. If G is the matrix in Exercise 2 and $g = (1, 2)^T$, solve the system (9.3.1) for $\lambda = 0.9, 0.8, 0.7$ and plot the resulting values of $||p||_2$.

9.4.   Results with the Newton method

We use NMp and NMw, respectively, to denote the OPTIMA implementations of the Newton method with perfect and weak line searches.

In the examples below, exact first derivatives are obtained using automatic differentiation but the Hessian is approximated by finite differences.

When the Hessian is not positive-definite, a search direction is obtained using a search direction derived from (9.3.4). The trial values of μ are given by

$$\mu_1 = \frac{0.01}{||G||_1}, \quad \mu_i = \mu_1 + \frac{0.33i}{||G||_1} \quad \text{for} \quad i = 2, \ldots, 6.$$

This not a particularly sophisticated implementation of the trust region approach but it does provide a reasonable safeguard against breakdown of the Newton algorithm.

The entries in Tables 9.1 and 9.2 are numbers of iterations and function calls needed by NMp and NMw to satisfy the stopping rule (4.3.2) to low, standard and high accuracy. (The figure in brackets is the number of iterations encountering a non positive-definite Hessian.)

Problem	Low Accuracy itns/fns	Standard Accuracy itns/fns	High Accuracy itns/fns
TD1	3/13	3/13	3/13
TD2	3/12	3/12	3/12
VLS1	1/2	1/2	1/2
TLS1	4(1)/41	4(1)/41	4(1)/41
VLS2	5(2)/23	5(2)/23	6(2)/24
R1(1)	5/35	5/35	6/37
R1(2)	15(15)/84**	22(19)/113	22(19)/113
OC1(4)	1/2	1/2	1/2
OC2(4)	3/13	4/14	4/14

Table 9.1. NMp solutions for Problems TD1–OC2.

Problem	Low Accuracy itns/fns	Standard Accuracy itns/fns	High Accuracy itns/fns
TD1	5/6	5/6	5/6
TD2	4/5	4/5	4/5
VLS1	1/2	1/2	1/2
TLS1	6(1)/8	6(1)/8	7(1)/9
VLS2	13(9)/14	13(9)/14	13(9)/14
R1(1)	9(1)/27	10(1)/28	10(1)/28
R1(2)	14(14)/27**	18(16)/32	19(16)/33
OC1(4)	1/2	1/2	1/2
OC2(4)	3/5	4/6	4/6

Table 9.2. NMw solutions for Problems TD1–OC2.

Some conclusions to be drawn from Tables 9.1, 9.2 are as follows.

• NMp and NMw converge in just one iteration on Problems VLS1 and OC1 which both have quadratic objective functions.
• On all the problems, the Newton method converges in fewer iterations and function calls than steepest descent (see Tables 7.1, 8.1) and direct search methods (Table 5.1).
• On the nonquadratic problems, NMw is typically more economical than NMp in terms of function evaluations even when it requires more iterations.
• The superscript "**" indicates that premature termination occurs on Problem R1(2) when the low-accuracy convergence test is used. The region round the solution is rather flat and the gradient norm becomes less than $10^{-4}\sqrt{n}$ while the search is still in a nonconvex region. Such premature convergence is always a risk when a stopping rule is based only on the gradient norm.

The Newton method has to deal with non positive-definite Hessians during some of the solutions reported above. On Problem TLS1, for instance, both NMp and NMw encounter an indefinite Hessian on the first iteration. However, the recovery procedure using (9.3.4) generates a suitable descent direction and the method goes on to converge to the correct solution. On problem R1(2) both NMp and NMw remain in a non-convex region for the first 20 iterations or so and the standard Newton correction is only used on the last few steps.

The entries across the three columns of Tables 9.1 and 9.2 demonstrate the practical implications of the Newton method's theoretical quadratic convergence rate. It quite often happens that a single iteration can take $\|\nabla F\|$ from failing the low-accuracy test (4.3.2) with $\epsilon = 10^{-4}$ to passing it with $\epsilon = 10^{-5}$ (standard accuracy). In such cases we see the same figures in two or even three of the columns in the table. It rarely takes more than one iteration to improve the accuracy of a solution by an order of magnitude. This can be contrasted with the much slower ultimate convergence of steepest descent implied by corresponding figures in Tables 7.1 and 8.1.

Exercises

1. Modify Problem TLS1 to find the best straight-line approximation to the points $(1,3)$, $(1.1, 3.2)$, $(1.3, 4)$, $(1.6, 4.7)$, $(1.9, 5.7)$ and then find a solution using the Newton method.
2. Modify Problem VLS2 to find an approximation to the data points
$(0,1)$, $(0.2, 0.95)$, $(0.4, 0.85)$, $(0.6, 0.65)$, $(0.8, 0.35)$

using the model $y = \cos(ax + b)$. Solve this problem by Newton's method.

3. Use NMp and NMw to solve Problems R1(1) and R1(2) with $\rho = 0.05$ and $\rho = 0.2$. Comment on the changes in the computed solutions and also in the numbers of iterations and function calls required.

4. Apply NMp and NMw to Problems OC2(6) and OC2(10).

5. Apply NMp and NMw to a modified form of Problem OC2(6) in which the initial and terminal conditions are

$$t_f = 2, \quad \tau = \frac{t_f}{n}, \quad s_0 = 0, \quad s_f = 0.9, \quad u_0 = 0, \quad u_f = 0.5.$$

6. By modifying an OPTIMA procedure, or otherwise, implement a version of the Newton method which reverts to the steepest descent method on any iteration where the Hessian is not positive-definite. Investigate how this method performs on Problem R1(2).

9.5. The Gauss–Newton method

Some of the problems TD1–OC2 have objective functions which are sums of squared terms. The least-squares data-fitting problems are obvious examples, but Problems OC1 and OC2 are also in this form. There is a variant of Newton's method for the special case of minimizing $F(x)$ when

$$F(x) = \sum_{i=1}^{m} f_i(x)^2 \tag{9.5.1}$$

(where we *assume* $m \geq n$). Differentiating (9.5.1) gives

$$\nabla F(x) = 2 \left\{ \sum_{i=1}^{m} \nabla f_i(x) f_i(x) \right\}.$$

If f is the m-vector whose elements are the *subfunctions* $f_i(x)$ and if J is the $m \times n$ *Jacobian* matrix whose ith row is $\nabla f_i(x)^T$ then we can also write

$$\nabla F(x) = 2J^T f. \tag{9.5.2}$$

Differentiating a second time gives

$$\nabla^2 F(x) = 2 \left\{ J^T J + \sum_{i=1}^{m} \nabla^2 f_i(x) f_i(x) \right\}. \tag{9.5.3}$$

In data-fitting problems, the subfunctions are often close to zero at a solution. It may also happen that the model function is chosen so that the f_i are nearly linear and hence $||\nabla^2 f_i(x)||$ is close to zero. In both situations the second term on the right hand side of (9.5.3) will be small in comparison with the first term. If we assume we can ignore this second term then $2J^T J$ becomes a convenient approximation to $\nabla^2 F$. This leads to an algorithm which resembles the Newton method but uses no second derivatives.

The Gauss–Newton method for minimizing a sum of squares

Choose x_0 as an estimate of x^*
Repeat for $k = 0, 1, 2, \ldots$
Set $f_k =$ the vector with elements $f_i(x_k)$
Set J_k as the corresponding Jacobian matrix
Obtain p_k by solving

$$(J_k^T J_k)p_k = -J_k^T f_k \qquad (9.5.4)$$

Find s so $F(x_k + sp_k)$ satisfies Wolfe conditions 2 and 3
Set $x_{k+1} = x_k + sp_k$
until $||J_k^T f_k||$ is sufficiently small.

The vector p_k used in this algorithm approximates the Newton direction because $2J_k^T f_k = \nabla F(x_k)$ and $2J_k^T J_k \approx \nabla^2 F(x_k)$. Because $J_k^T J_k$ can be shown to be positive semi-definite we can (fairly) safely assume that p_k is a descent direction, satisfying Wolfe condition 1. We refer to the system (9.5.4) which gives the Gauss–Newton search direction as the *normal equations*.

The Gauss–Newton algorithm can often minimize a function of the form (9.5.1) in fewer iterations than more general unconstrained optimization methods. However, it may do more work per iteration than a Newton method because $O(n^2 m) + O(\frac{1}{6}n^3)$ multiplications are needed to form $J_k^T J_k$ and then factorize it by the Cholesky method. The Gauss–Newton method will have a cost advantage if the calculation of $J_k^T J_k$ is less expensive than the evaluation of the full Hessian $\nabla^2 F(x_k)$.

The Gauss–Newton algorithm given above will fail in the exceptional case that $J_k^T J_k$ is singular. However, if we choose some $\lambda > 0$ we can obtain a downhill search direction, p_k, from the *Levenberg–Marquardt* equations [44, 47]

$$(J_k^T J_k + \lambda I)p_k = -J_k^T f_k. \qquad (9.5.5)$$

As explained in section 9.3, this search direction minimizes a quadratic model of F subject to a limit on the Euclidian norm of p. In other words it solves

$$\text{Minimize} \quad \frac{1}{2}(p^T J_k^T J_k p) + p^T J_k^T f_k \quad \text{subject to} \quad ||p||_2 \leq \Delta$$

for some positive Δ. The relationship between λ and Δ is not simple, but we can easily see that, as $\lambda \to \infty$, p_k tends towards an infinitesimal step along the steepest descent direction $-J_k^T f_k$.

Exercises

1. Write down expressions for the subfunctions and the elements of the Jacobian matrix for Problems VLS1 and TLS1.
2. Show that the matrix $J_k^T J_k$ is at least positive-semi-definite. Show also that it is positive-definite if the columns of J_k are linearly independent.

9.6.   Results with the Gauss–Newton method

We use GNp and GNw to denote the OPTIMA implementations of the Gauss–Newton method with perfect and weak line searches. In the examples below, exact first derivatives of the function and subfunctions are obtained using automatic differentiation. In the case when $J_k^T J_k$ is singular, the search direction is computed from (9.5.5), using a similar approach to that based on (9.3.4).

The entries in Tables 9.3 and 9.4 show the numbers of iterations and function calls needed to satisfy the stopping rule (4.3.2) with different values of ϵ.

Problem	Low Accuracy itns/fns	Standard Accuracy itns/fns	High Accuracy itns/fns
VLS1	1/2	1/2	1/2
TLS1	3/13	4/14	4/14
VLS2	8/125	9/127	10/129
OC1(4)	1/2	1/2	1/2
OC2(4)	3/13	5/14	7/18

Table 9.3.   GNp solutions for Problems VLS1–OC2.

Problem	Low Accuracy itns/fns	Standard Accuracy itns/fns	High Accuracy itns/fns
VLS1	1/2	1/2	1/2
TLS1	5/7	5/7	6/8
VLS2	8/9	9/10	10/11
OC1(4)	1/2	1/2	1/2
OC2(4)	6/8	10/12	14/16

Table 9.4.   GNw solutions for Problems VLS1–OC2.

From the results in the tables we can make the following observations.

• As with the Newton method, Gauss–Newton converges in one iteration when F is quadratic (Problems VLS1, OC1).

• The Gauss–Newton method can outperform the Newton approach, especially with weak line searches. However it does much less well than Newton's method on Problem VLS2 when a perfect search is used. This may be due to the fact that residuals of the data-fitting Problem VLS2 are not zero and therefore $J^T J$ is not so good an approximation to G as it is for TLS1 and VLS1.

• The fallback search direction (9.5.5) is never used in any of the quoted solutions. In nonconvex regions the approximation $2J_k^T J_k$ has an advantage over the true Hessian in that the normal equations (9.5.4) almost invariably yield a descent direction even when the Newton correction is uphill.

• Ultimate convergence of the Gauss–Newton method does not seem to be as fast as that for the Newton method. On Problem OC2, for instance, it can take more than one iteration to improve solution accuracy by an order of magnitude.

Exercises

1. Use the Gauss–Newton method to solve the first question in the exercises of section 9.4.

2. Use the Gauss–Newton method to solve the second question in the exercises of section 9.4.

3. Use GNp and GNw to solve Problems OC1(4) and OC2(4) when $\rho = 0.05$ and $\rho = 0.2$. Comment on the results.

4. Use results from Tables 9.3 and 9.4 to estimate the rate of convergence of GNp and GNw on Problem OC2(4). Extend your investigation to the problem OC2(8).

5. Apply GNp and GNw to a modified form of Problem VLS2 in which the model function is

$$z = \phi(t, x) = x_1 e^{x_2 t} + x_3.$$

Chapter 10

Quasi-Newton Methods

10.1.   Approximate second-derivative information

Drawbacks of the Newton method were noted in Section 9.2. These have led to the development of *quasi-Newton* techniques (sometimes called *variable-metric methods*). The essential idea of these methods is simply that a positive-definite matrix is used to approximate the Hessian (or its inverse). This saves the work of computing exact second derivatives and also avoids the difficulties associated with loss of positive-definiteness. The approximating matrix is updated on each iteration so that, as the search proceeds, second derivative information is improved. Before going into detail about this updating we give an outline quasi-Newton algorithm.

An outline quasi-Newton method

Choose x_0 as an initial estimate of the minimum of $F(x)$.
Choose H_0 as an arbitrary symmetric positive definite matrix
Repeat for $k = 0, 1, 2, \ldots$
Set $g_k = \nabla F(x_k)$
Set $p_k = -H_k g_k$
Find s, so $F(x_k + sp_k)$ satisfies (8.1.3), (8.1.4) for some η_1, η_2
set $x_{k+1} = x_k + sp_k, \quad \gamma_k = g_{k+1} - g_k, \quad \delta_k = x_{k+1} - x_k$
Obtain a new positive definite matrix H_{k+1} such that

$$H_{k+1}\gamma_k = \delta_k \tag{10.1.1}$$

until $\|\nabla F(x_{k+1})\|$ is sufficiently small.

M. Bartholomew-Biggs, *Nonlinear Optimization with Engineering Applications*,
DOI: 10.1007/978-0-387-78723-7_10, © Springer Science+Business Media, LLC 2008

In this algorithm, H_k is an estimate of the inverse Hessian $\nabla^2 F(x_k)^{-1}$. The simple initial choice $H_0 = I$, the identity matrix, is usually satisfactory.

Definition The equation (10.1.1) used in the calculation of the new matrix H_{k+1} is called the *quasi-Newton condition*.

Condition (10.1.1) is derived as follows. If $F(x) = \frac{1}{2}x^T A x + b^T x + c$ then

$$\gamma_k = g_{k+1} - g_k = (Ax_{k+1} + b) - (Ax_k + b) = A(x_{k+1} - x_k) = A\delta_k.$$

In other words

$$A^{-1}\gamma_k = \delta_k.$$

Thus, when $F(x)$ is a quadratic function, the condition (10.1.1) causes H_{k+1} to share a property with the true inverse Hessian.

To save computing effort – and also to preserve second-derivative information already present in H_k – the new matrix H_{k+1} is obtained by a low-rank modification to H_k. This means that H_{k+1} is of the form

$$H_{k+1} = H_k + auu^T \quad \text{or} \quad H_{k+1} = H_k + buu^T + cvv^T$$

where a, b, c are scalars and u, v are vectors depending on H_k, γ_k and δ_k. We now describe some widely-used *updating formulae*.

10.2.   Rank-two updates for the inverse Hessian

Definition The *Davidon–Fletcher–Powell* (DFP) update [15, 22] for H_{k+1} is

$$H_{k+1} = H_k - \frac{H_k\gamma_k\gamma_k^T H_k}{\gamma_k^T H_k\gamma_k} + \frac{\delta_k\delta_k^T}{\delta_k^T\gamma_k}. \tag{10.2.1}$$

Proposition The DFP formula makes H_{k+1} satisfy (10.1.1).

Proof The result follows immediately on multiplying the right-hand side of (10.2.1) by γ_k and simplifying.

Proposition   The DFP formula causes H_{k+1} to inherit positive-definiteness from H_k provided

$$\delta_k^T\gamma_k > 0. \tag{10.2.2}$$

(The proof of this is left to the reader – see Exercise 3, below.)

The quasi-Newton condition does not define H_{k+1} uniquely because it consists of n equations involving the n^2 elements of H_{k+1}. As well as the

DFP formula, the *Broyden–Fletcher–Goldfarb–Shanno* (BFGS) formula [12] also causes H_{k+1} to satisfy (10.1.1).

Definition The BFGS formula for H_{k+1} is

$$H_{k+1} = H_k - \frac{H_k\gamma_k\delta_k^T + \delta_k\gamma_k^T H_k}{\delta_k^T\gamma_k} + \left[1 + \frac{\gamma_k^T H_k\gamma_k}{\delta_k^T\gamma_k}\right]\frac{\delta_k\delta_k^T}{\delta_k^T\gamma_k}. \quad (10.2.3)$$

This formula also ensures that H_{k+1} is positive-definite when (10.2.2) holds.

An important result which links the DFP and BFGS updates is the following.

Dixon's Theorem [18, 19] If a quasi-Newton algorithm includes a perfect line search then, for any function $F(x)$, the same sequence of iterates $\{x_k\}$ will be obtained irrespective of whether H_k is produced by the DFP or BFGS formula.

This theorem seems to imply there is no practical difference between the DFP and BFGS updates. However, when we attempt a perfect line search in finite-precision arithmetic, rounding errors can prevent the condition $p_k^T g_{k+1} = 0$ from being satisfied precisely. It turns out that even small departures from "perfection" in the line search can cause differences to appear in the iterates given by different updates. Moreover, most quasi-Newton implementations use weak line searches and then Dixon's theorem does not apply.

In practice, the BFGS update is usually preferred to the DFP one. Experience suggests that, although both (10.2.1) and (10.2.3) keep H_k positive-definite, the DFP formula is more likely to produce matrices which are near-singular and this can have an adverse affect on its performance.

An example

We consider a quasi-Newton iteration (with perfect line search and DFP update) on the function

$$F(x) = x_1^2 + 3x_2^2 + x_1x_2 + x_1 + x_2.$$

We dispense with iteration-number subscripts for this example and take the starting point as $x = (0, 0)^T$. To make the iteration differ from steepest descent we use the initial inverse Hessian estimate

$$H = \begin{pmatrix} \frac{1}{2} & 0 \\ 0 & \frac{1}{6} \end{pmatrix}.$$

The gradient of $F(x)$ is

$$g = \begin{pmatrix} 2x_1 + x_2 + 1 \\ 6x_2 + x_1 + 1 \end{pmatrix}$$

and therefore the initial search direction is

$$p = -Hg = -\begin{pmatrix} \frac{1}{2} & 0 \\ 0 & \frac{1}{6} \end{pmatrix}\begin{pmatrix} 1 \\ 1 \end{pmatrix} = \begin{pmatrix} -\frac{1}{2} \\ -\frac{1}{6} \end{pmatrix}.$$

The new point is

$$x^+ = x + sp = \left(-\frac{s}{2}, -\frac{s}{6}\right)^T$$

where s is chosen to minimize

$$F(x + sp) = \frac{s^2}{4} + \frac{3s^2}{36} + \frac{s^2}{12} - \frac{s}{2} - \frac{s}{6} = \frac{5s^2}{12} - \frac{2s}{3}.$$

This gives $s^* = 0.8$ and so the new point and new gradient are

$$x^+ = \begin{pmatrix} -0.4 \\ -0.1333 \end{pmatrix} \quad \text{and} \quad g^+ = \begin{pmatrix} 0.0667 \\ -0.2 \end{pmatrix}.$$

Now, using quasi-Newton notation,

$$\delta = x^+ - x = \begin{pmatrix} -0.4 \\ -0.1333 \end{pmatrix}, \quad \gamma = g^+ - g = \begin{pmatrix} -0.9333 \\ -1.2 \end{pmatrix}.$$

Thus, working to five significant figures, $\delta^T\gamma = 0.53328$. Moreover,

$$H\gamma = \begin{pmatrix} -0.46665 \\ -0.2 \end{pmatrix}$$

and so $\gamma^T H\gamma = 0.67552$. We also obtain

$$H\gamma\gamma^T H = \begin{pmatrix} 0.21776 & 0.09333 \\ 0.09333 & 0.04 \end{pmatrix}$$

so that $\quad \dfrac{H\gamma\gamma^T H}{\gamma H\gamma} = \begin{pmatrix} 0.32236 & 0.13816 \\ 0.13816 & 0.05921 \end{pmatrix}$

and

$$\delta\delta^T = \begin{pmatrix} 0.16 & 0.05332 \\ 0.05332 & 0.017769 \end{pmatrix} \quad \text{so that} \quad \frac{\delta\delta^T}{\delta^T\gamma} = \begin{pmatrix} 0.3 & 0.1 \\ 0.1 & 0.03333 \end{pmatrix}.$$

Putting these ingredients together in the DFP formula,

$$H^+ = \begin{pmatrix} 0.5 - 0.32236 + 0.3 & 0 - 0.13816 - 0.1 \\ 0 - 0.13816 - 0.1 & 0.16667 - 0.05921 + 0.03333 \end{pmatrix}$$

$$= \begin{pmatrix} 0.47764 & -0.03816 \\ -0.03816 & 0.14079 \end{pmatrix}.$$

On the next iteration, the search direction will be

$$p = -H^+ g^+ = \begin{pmatrix} -0.039474 \\ 0.030702 \end{pmatrix}.$$

The reader can verify that the perfect step $s^* \approx 1.3807$ along p away from x^+ will locate the minimum of $F(x)$ at $(-\frac{5}{11}, -\frac{1}{11})^T$ (subject to rounding errors in five-digit arithmetic).

Exercises
1. For the worked example above, does the second iteration locate the solution if H^+ is obtained by the BFGS update?
2. Using the DFP update and perfect line searches, do two quasi-Newton iterations on the function

$$F(x) = x_1^2 + x_1 x_2 + \frac{x_2^2}{2}$$

   starting from $x = (1,1)$. What happens if a weak line search is used instead?
3. Prove that the DFP update ensures that H_{k+1} inherits positive-definiteness from H_k provided $\delta_k^T \gamma_k > 0$.
   (*Hints: (i)* a positive-definite H_k has a Cholesky factor L such that $H_k = LL^T$;
   (ii) if u and v are vectors the Schwarz inequality states that $(u^T v)^2 \le u^T u \, v^T v$.)
4. Prove that the condition (10.2.2) for ensuring positive-definiteness in DFP and BFGS updates is automatically satisfied when a perfect line search is used.
5. If F is a quadratic function and if $H_k = (\nabla^2 F(x_k))^{-1}$ show that H_{k+1} given by the DFP update is equal to $(\nabla^2 F(x_{k+1}))^{-1}$. Is the same true if H_{k+1} is given by the BFGS update?
6. Show that the following general result follows from Dixon's theorem. If a quasi-Newton algorithm includes a perfect line search then, for any function $F(x)$, the same sequence of iterates $\{x_k\}$ will be produced when the update for H_{k+1} is any member of the family defined by

$$H_{k+1} = \theta H_{k+1}^{dfp} + (1-\theta) H_{k+1}^{bfgs} \qquad (10.2.4)$$

where $1 \geq \theta \geq 0$ and H_{k+1}^{dfp}, H_{k+1}^{bfgs} denote the right-hand sides of the updating formulae (10.2.1) and (10.2.3).

10.3.   Convergence of quasi-Newton methods

There are a number of convergence results about quasi-Newton methods based on the DFP and BFGS updates. The following propositions all assume exact arithmetic is used (i.e., there are no rounding errors).

Proposition   [12] If $F(x)$ is an n-variable convex quadratic function then a quasi-Newton algorithm, with perfect line search, will converge to the minimum of F in at most n iterations with both the DFP and BFGS update. Moreover $H_n = \nabla^2 F^{-1}$.

Proposition   (Powell [54]) If $F(x)$ is a twice-differentiable function which is convex in some region R around a local minimum x^*, then a quasi-Newton algorithm, with perfect line search and either the DFP or BFGS update, will converge to x^* from any starting point in R.

Proposition   (Powell [55]) If $F(x)$ is a twice-differentiable function which is convex in some region R around a local minimum x^*, then a quasi-Newton algorithm, with a weak line search and the BFGS update, will converge to x^* from any starting point in R.
(A similar result about convergence of a quasi-Newton algorithm with a weak line search and the DFP update has also been proved [56] but stronger conditions on steplength are needed than for the BFGS version. This may help to explain the generally observed practical superiority of the BFGS version.)

Because they do not use the exact Hessian, quasi-Newton methods do not usually converge as quickly as the Newton method. Performance near the solution is, however, superior to that of the steepest descent approach.

Proposition   [54] If H_k tends to the true inverse Hessian as x_k approaches x^* and if the stepsize $s = 1$ satisfies Wolfe conditions (8.1.3), (8.1.4) for all $k \geq K$ then quasi-Newton methods are capable of ultimately *superlinear* convergence. This means that, for k sufficiently large,

$$\frac{||x_{k+1} - x^*||}{||x_k - x^*||} \to 0$$

or, equivalently, that the error norm decreases at a rate implied by

$$||x_{k+1} - x^*|| = C||x_k - x^*||^r$$

for some constant C and for $1 < r < 2$. This is not as good as the quadratic ($r = 2$) convergence given by the Newton method but it is superior to the linear ($r = 1$) convergence of the steepest descent algorithm.

Because the updating formulae for H_{k+1} involve only vector-vector and matrix-vector products, the number of multiplications per iteration of a quasi-Newton method varies with n^2. This compares favourably with the $O(n^3)$ multiplications per iteration needed by the Newton method to form and factorize the Hessian $\nabla^2 F(x)$. On the other hand, the Newton method may take significantly fewer iterations and so there is not always a clear-cut advantage in runtime for quasi-Newton methods.

10.4. Results with quasi-Newton methods

The OPTIMA implementations of the quasi-Newton approach are called QNp and QNw to denote the use of a perfect or a weak line search. In both cases the BFGS updating formula (10.2.3) is used. Tables 10.1 and 10.2 show numbers of iterations and function calls needed to solve Problems TD1–OC2. As in previous chapters, we quote results for three levels of convergence accuracy.

Problem	Low Accuracy itns/fns	Standard Accuracy itns/fns	High Accuracy itns/fns
TD1	4/27	5/29	5/29
TD2	4/18	5/20	5/20
VLS1	2/5	2/5	2/5
TLS1	4/21	5/23	5/23
VLS2	3/43**	11/229	11/229
R1(1)	6/37	6/37	7/40
R1(2)	12/69	12/69	13/71
OC1(4)	3/7	4/10	4/10
OC2(4)	6/19	7/21	7/21

Table 10.1. QNp solutions for Problems TD1–OC2.

Noteworthy points about Tables 10.1 and 10.2 are as follows.

• Convergence of QNp on Problems VLS1 and OC1(4) matches theoretical expectations in the first proposition of section 10.3. These are quadratic problems with $n = 2$ and $n = 4$, respectively, and a quasi-Newton method with perfect line search should converge in (at most) n iterations. Note that, although QNw needs more iterations than QNp, it uses fewer function calls per line search.

Problem	Low Accuracy itns/fns	Standard Accuracy itns/fns	High Accuracy itns/fns
TD1	8/12	10/14	10/14
TD2	7/10	8/11	9/12
VLS1	2/5	2/5	2/5
TLS1	30/92	31/93	32/94
VLS2	21/78	22/79	23/80
R1(1)	10/22	11/23	13/25
R1(2)	19/37	22/40	22/40
OC1(4)	7/9	9/11	13/15
OC2(4)	8/10	9/11	9/11

Table 10.2. QNw solutions for Problems TD1–OC2.

• On the nonquadratic problems, QNw typically uses more iterations than QNp. In terms of overall workload, this is sometimes outweighed by a decrease in function calls per iteration of QNw. An exception occurs on Problem TLS1 where QNp is much more efficient than QNw.

• The low-accuracy result by QNp for Problem VLS2 is anomalous. The low tolerance on the gradient norm causes the search to stop prematurely at a point which is not close to the true minimum. This is not a failing of the quasi-Newton approach itself, but rather a warning that any iterative technique can give misleading results if convergence tests are not strict enough.

• Comparison with Tables 7.1, 8.1, 9.1 and 9.2 shows that the quasi-Newton approach is quite competitive with the Newton method and is considerably more efficient than steepest descent.

• Unlike the quadratically convergent Newton method, which seldom needs more than one iteration to go from low to high accuracy, quasi-Newton methods are only capable of superlinear convergence. As a consequence, it is quite common for both QNp and QNw to take two iterations to reduce the gradient norm from $O(10^{-4})$ to $O(10^{-6})$. This performance is, however, much better than that of the linearly convergent steepest descent method in Tables 7.1 and 8.1. (We no longer consider the steepest descent method as a serious contender for solving practical problems.)

Exercises

1. Use QNp and QNw to solve problem 1 from the exercises in Section 9.4.
2. Use QNp and QNw to solve problem 2 from the exercises in section 9.4.
3. Investigate the performance of quasi-Newton methods when applied to Problems R1(1) and R1(2) as ρ increases.
4. Print out (and, if possible, plot) the iterates obtained by QNw applied to Problem TLS1.

5. Compare the performance of QNw and NMw on Problems OC2(8) and OC2(10).

6. Apply QNp to a modified form of Problem TD2 which involves a closed tank and has a target surface area $S^* = 40$.

7. Implement a quasi-Newton algorithm which uses the DFP update instead of the BFGS formula. Test its performance on Problems TD1–OC2.

10.5. Some further updating formulae

The DFP and BFGS formulae change the matrix H_k in the two-dimensional subspace spanned by δ_k and $H_k\gamma_k$. The *symmetric rank-one* (SR1) formula, however, only alters H_k in the one-dimensional space spanned by the vector $(\delta_k - H_k\gamma_k)$.

Definition The *symmetric rank one* updating formula [26] is

$$H_{k+1} = H_k + \frac{v_k v_k^T}{v_k^T \gamma_k} \quad \text{where} \quad v_k = (\delta_k - H_k\gamma_k). \tag{10.5.1}$$

It is easy to show that (10.5.1) satisfies the quasi-Newton condition. In fact it is the only symmetric rank-one update which will do so.

The update (10.5.1) has an interesting "memory property" when used with quadratic functions.

Proposition If $H_k\gamma_{k-1} = \delta_{k-1}$ (and so H_k agrees with the true inverse Hessian for the vectors γ_{k-1} and δ_{k-1}) then H_{k+1} given by (10.5.1) satisfies

$$H_{k+1}\gamma_k = \delta_k \quad \text{and} \quad H_{k+1}\gamma_{k-1} = \delta_{k-1}.$$

Proof of this property is left to the reader. From it there follows

Proposition If $F(x)$ is an n-variable convex quadratic function then a quasi-Newton algorithm, using a weak line search and the SR1 update will obtain $H_n = \nabla^2 F^{-1}$, and therefore will converge to the minimum of F in at most $n + 1$ iterations.

In one sense, SR1 is better than DFP or BFGS because it gives finite termination on a quadratic function without perfect line searches. However (10.5.1) has the drawback that it may not keep H_{k+1} positive definite under the same, easy to check, condition (10.2.2) as applies to the DFP or BFGS formulae. Indeed even when SR1 is used on a positive-definite quadratic function some of the intermediate H_k may be indefinite.

Exercises

1. Prove that (10.5.1) satisfies the quasi-Newton condition and show that there is no other suitable update of the form $H_{k+1} = H_k + \alpha vv^T$.

2. Prove the "memory property" of the symmetric rank-one update, namely:
   If $H_k \gamma_{k-1} = \delta_{k-1}$ then the matrix H_{k+1} given by (10.5.1) satisfies

   $$H_{k+1}\gamma_k = \delta_k \quad \text{and} \quad H_{k+1}\gamma_{k-1} = \delta_{k-1}.$$

3. Do two quasi-Newton iterations with weak search and SR1 update on

   $$F(x) = x_1^2 + x_1 x_2 + \frac{x_2^2}{2}$$

   starting from $x = (1,1)$. Comment on the outcome.

4. Investigate conditions which will ensure that H_{k+1} given by the SR1 update will inherit positive definiteness from H_k.

5. Implement a quasi-Newton procedure which uses the SR1 update. How does it perform on problems considered in the previous section? (Your answer should deal with both perfect and weak line searches.)

Updating estimates of the Hessian

Some implementations of the quasi-Newton technique work with estimates, B_k, of $\nabla^2 F$ rather than $\nabla^2 F^{-1}$. (It can be argued that approximating the Hessian is a more numerically stable process than approximating its inverse.) The quasi-Newton condition for B_{k+1} is, of course,

$$B_{k+1}\delta_k = \gamma_k. \tag{10.5.2}$$

It can be shown that the DFP and BFGS formulae are *dual* in the following sense. If $H_k = (B_k)^{-1}$ then the update which gives $B_{k+1} = (H_{k+1}^{bfgs})^{-1}$ is

$$B_{k+1} = B_k - \frac{B_k \delta_k \delta_k^T B_k}{\delta_k^T B_k \delta_k} + \frac{\gamma_k \gamma_k^T}{\delta_k^T \gamma_k}. \tag{10.5.3}$$

This is precisely the DFP formula with B replacing H and with δ and γ interchanged. Similarly $B_{k+1} = (H_{k+1}^{dfp})^{-1}$ is found by replacing H with B and exchanging δ and γ in the BFGS update.

The SR1 formula (10.5.1) is *self-dual* because $B_{k+1} = (H_{k+1}^{sr1})^{-1}$ is given by

$$B_{k+1} = B_k + \frac{w_k w_k^T}{w_k^T \delta_k} \quad \text{where} \quad w_k = (\gamma_k - B_k \delta_k). \tag{10.5.4}$$

It might seem inefficient in practice to use an algorithm involving B instead of H because p_k will then be obtained from

$$B_k p_k = -g_k \qquad (10.5.5)$$

which implies that B_k must be factorized. This factorization cost is avoided when H_k approximates the inverse Hessian. However, Gill and Murray [30] have shown it is possible to store and update the Cholesky factors of B_k. This makes it much more economical to solve (10.5.5) on every iteration.

Much more work has been done on the theory and implementation of quasi-Newton methods than can be contained in a single chapter. For fuller accounts of other updating formulae and algorithms see [26] and [17].

Exercises

1. If a nonsingular matrix Q is updated to become $\tilde{Q} = Q + uu^T$ show that

$$\tilde{Q}^{-1} = Q^{-1} - \frac{Q^{-1}uu^T Q^{-1}}{1 + u^T Q^{-1} u}. \qquad (10.5.6)$$

   (This is called the *Sherman–Morrison–Woodbury* formula.)

2. Use (10.5.6) to show that if $H_k = B_k^{-1}$ and if B_{k+1} is given by (10.5.4) then $H_{k+1}^{sr1} = B_{k+1}^{-1}$.

Loss adjusters (Part 1) [5]

They walk beside disused canals
wearing matching jackets. At the collars
slightly shiny uncut hair
has curled, untidy as an unkept promise.
Afterwards, behind uncurtained windows,
they resume a sleepless dialogue
on lists of post-disaster redesigns.

A strain-gauge to tell if the building is bulging;
foundations dug deeper to shore up the spire;
conventional spars should replace surface bracing;
make fuel-chamber gaskets resistant to fire.

Chapter 11

Conjugate Gradient Methods

11.1. Conjugate gradients for a quadratic $Q(x)$

We have already shown that the minimum of a convex quadratic function

$$Q(x) = \frac{1}{2}(x^T A x) + b^T x + c$$

can be found by solving $\nabla Q = 0$ which is equivalent to $Ax = -b$. When A is symmetric and positive definite, the system $Ax = -b$ can be solved by an iterative technique called the *conjugate gradient method* [35]. The theory behind this method is based on the following definition.

Definition Two vectors u and v are said to be *conjugate* with respect to a symmetric matrix A if

$$u^T A v = 0. \tag{11.1.1}$$

Conjugate gradient method for solving $Ax = -b$

Choose x_0 as an initial estimate of the solution
Calculate $g_0 = Ax_0 + b$. Set $p_0 = -g_0$
Repeat for $k = 0, 1, 2, \ldots$
find s so that $p_k^T g_{k+1} = p_k^T (A(x_k + sp_k) + b) = 0$
set $x_{k+1} = x_k + sp_k$
determine β and p_{k+1} using

$$\beta = \frac{g_{k+1}^T g_{k+1}}{g_k^T g_k} \quad \text{and} \quad p_{k+1} = -g_{k+1} + \beta p_k \tag{11.1.2}$$

until $\|g_{k+1}\|$ is sufficiently small.

M. Bartholomew-Biggs, *Nonlinear Optimization with Engineering Applications*,
DOI: 10.1007/978-0-387-78723-7_11, © Springer Science+Business Media, LLC 2008

The step, s, along the search direction p_k in the algorithm is given by

$$s = -\frac{p_k^T g_k}{p_k^T A p_k}. \qquad (11.1.3)$$

This gives $p_k^T g_{k+1} = 0$, and is equivalent to choosing s to minimize $Q(x_k + s p_k)$.

The formula (11.1.2) for calculating β is designed to make the search directions conjugate with respect to A, that is

$$p_i^T A p_j = 0 \quad \text{when } i \neq j. \qquad (11.1.4)$$

For the moment we simply state (11.1.4) as a fact and show how it motivates the conjugate gradient algorithm. We consider the justification of (11.1.4) in a later section.

To show the significance of making the search directions mutually conjugate with respect to A, we first state and prove a result involving the first two iterations of the conjugate gradient algorithm.

Proposition After two iterations of the conjugate gradient method, the gradient $g_2 = A x_2 + b$ satisfies

$$p_1^T g_2 = p_0^T g_2 = 0. \qquad (11.1.5)$$

Proof After the first iteration the new point is x_1 and so $g_1 = A x_1 + b$. Because of the perfect line search we also have $p_0^T g_1 = 0$.
Now consider iteration two. It will generate a point

$$x_2 = x_1 + s p_1 \quad \text{where} \quad g_2 = A x_2 + b \quad \text{and} \quad p_1^T g_2 = 0.$$

To prove the second part of (11.1.5) we note that

$$p_0^T g_2 = p_0^T (A x_1 + s A p_1 + b) = p_0^T g_1 + s p_0^T A p_1.$$

The first term in the rightmost expression is zero because of the line search on iteration one. The second is zero because p_0 and p_1 are conjugate w.r.t. A. Hence (11.1.5) holds.

This result means that the gradient after two iterations is orthogonal to both search directions p_0 and p_1. Similarly, we can prove a more general result.

Proposition After k iterations of the conjugate gradient method the gradient $g_k = A x_k + b$ satisfies

$$p_j^T g_k = 0 \quad \text{for } j = 0, 1, 2, \ldots, k-1. \qquad (11.1.6)$$

This proposition implies that, after k iterations, the gradient g_k is restricted to the $(n-k)$-dimensional subspace orthogonal to the vectors $p_0, \ldots, p_{k-1}$. From this we can deduce an important *finite termination property*.

Proposition The conjugate gradient method solves an $n \times n$ system $Ax = -b$ in at most n iterations.

Proof Property (11.1.6) implies that, after n iterations, g_n is orthogonal to the n vectors $p_0, p_1, \ldots, p_{n-1}$. But this means that it must lie in a subspace of dimension zero and so $g_n = 0$ which implies $Ax_n = -b$.

This finite termination property is only guaranteed for calculations involving exact arithmetic. In practice, (11.1.5), (11.1.6) may not be satisfied exactly when the iterations are performed in real arithmetic which is subject to rounding errors. Hence (a few) more than n conjugate gradient iterations may be needed for convergence to the solution of some $n \times n$ systems.

A worked example

We apply the conjugate gradient method to the function

$$f(x) = x_1^2 + x_1 x_2 + \frac{x_2^2}{2}$$

starting from $x_0 = (1, 1)^T$. The gradient vector is

$$g = \begin{pmatrix} 2x_1 + x_2 \\ x_1 + x_2 \end{pmatrix}$$

and so the search direction away from x_0 is

$$p_0 = -g_0 = \begin{pmatrix} -3 \\ -2 \end{pmatrix}.$$

Hence the new point will be of the form $x_1 = (1 - 3s, \ 1 - 2s)^T$ where s is chosen so that $p_0^T g_1 = 0$, where

$$g_1 = \begin{pmatrix} 2 - 6s + 1 - 2s \\ 1 - 3s + 1 - 2s \end{pmatrix} = \begin{pmatrix} 3 - 8s \\ 2 - 5s \end{pmatrix}.$$

Hence

$$p_0^T g_1 = -3(3 - 8s) - 2(2 - 5s).$$

By solving $p_0^T g_1 = 0$ we get the perfect steplength and the new point as

$$s^* = \frac{13}{34} \quad \text{and} \quad x_1 = \frac{1}{34} \begin{pmatrix} -5 \\ 8 \end{pmatrix} \quad \text{where} \quad g_1 = \frac{1}{34} \begin{pmatrix} -2 \\ 3 \end{pmatrix}.$$

We now use g_1 to find a search direction for the next iteration. First we get

$$\beta = \frac{g_1^T g_1}{g_0^T g_0} = \frac{1}{34^2}$$

and then

$$p_1 = -g_1 + \beta p_0 = \frac{1}{34}\begin{pmatrix} 2 \\ -3 \end{pmatrix} + \frac{1}{34^2}\begin{pmatrix} -3 \\ -2 \end{pmatrix} = \frac{1}{34^2}\begin{pmatrix} 65 \\ -104 \end{pmatrix}.$$

The new solution estimate reached at the end of the second iteration will be

$$x_2 = x_1 + s p_1 = \left(-\frac{5}{34} + \frac{65s}{34^2}, \quad \frac{8}{34} - \frac{104s}{34^2} \right)^T$$

which gives

$$g_2 = \left(-\frac{2}{34} + \frac{26s}{34^2}, \quad \frac{3}{34} - \frac{39s}{34^2} \right)^T.$$

For a perfect line search the steplength s satisfies $p_1^T g_2 = 0$. This means

$$-\frac{130}{34^3} + \frac{(65 \times 26s)}{34^4} - \frac{312}{34^3} + \frac{(104 \times 39s)}{34^4} = 0.$$

After simplification this leads to

$$s^* = 34 \times \frac{442}{5746} \approx 2.5562.$$

Thus, after two iterations, the conjugate gradient method has reached

$$x_2 = \left(-\frac{5}{34} + \frac{(34 \times 442 \times 65)}{5746 \times 34^2}, \quad \frac{8}{34} - \frac{(34 \times 442 \times 104)}{5746 \times 34^2} \right)^T.$$

On simplification this gives $x_2 = (0, \ 0)^T$. This point minimizes the function because $g(x_2) = 0$. Hence the example demonstrates the finite termination property of the conjugate gradient method applied to a quadratic function.

Exercises

1. Do two conjugate gradient iterations, starting from $x = (0,0)^T$, applied to
$$F(x) = 2x_1^2 + x_1 x_2 + x_2^2 + x_1 - x_2.$$
   What do you observe about the result?

2. When the conjugate gradient algorithm is used to solve $g = Ax + b = 0$ show that the stepsize calculation (11.1.3) will ensure that

$p_k^T g_{k+1} = 0$. Show also that this value of s can be found, without using A directly, from

$$s = -\frac{p_k^T g_k}{p_k^T (g^+ - g_k)}$$

where $g^+ = Ax^+ + b$ and $x^+ = x_k + p_k$.

3. Extend the proof of (11.1.5) to prove (11.1.6).
4. Show that the eigenvectors of a symmetric matrix A are also conjugate directions with respect to A.
5. A quasi-Newton method with perfect line searches and using the DFP update is applied to a quadratic function F. Show that successive search directions are conjugate with respect to $\nabla^2 F$.

Conjugacy of search directions given by (11.1.2)

We now turn to a justification of the conjugacy property (11.1.4). The following propositions form part of a proof by induction.

Proposition The recurrence (11.1.2) ensures that $p_0^T A p_1 = 0$ and hence makes p_1 and p_0 conjugate w.r.t. A.

Proof We know the following:

$$p_0 = -g_0; \quad x_1 = x_0 + s p_0; \quad g_1 = g_0 + s A p_0 \qquad (11.1.7)$$

and by the perfect line search

$$p_0^T g_1 = 0 \quad \text{and} \quad s = -\frac{p_0^T g_0}{p_0^T A p_0}. \qquad (11.1.8)$$

We also have

$$\beta = \frac{g_1^T g_1}{g_0^T g_0} \quad \text{and} \quad p_1 = -g_1 + \beta p_0 \qquad (11.1.9)$$

From (11.1.9)

$$p_0^T A p_1 = -p_0^T A g_1 + \beta p_0^T A p_0$$

and from (11.1.7)

$$A p_0 = \frac{1}{s}(g_1 - g_0)$$

and so

$$p_0^T A p_1 = \frac{1}{s}(-g_1^T g_1 + g_0^T g_1 + \beta g_1^T p_0 - \beta g_0^T p_0).$$

But $p_0 = -g_0$ and so $g_0^T g_1 = -p_0^T g_1 = 0$ by (11.1.8). Therefore

$$p_0^T A p_1 = \frac{1}{s}(-g_1^T g_1 + \beta g_0^T g_0).$$

Now the definition of β (11.1.9) implies $p_0^T A p_1 = 0$.

Proposition The search direction calculation (11.1.2) implies $p_k^T g_k = -g_k^T g_k$

Proof We know that $p_k = -g_k + \beta p_{k-1}$ and so

$$p_k^T g_k = -g_k^T g_k + \beta p_{k-1}^T g_k.$$

But $p_{k-1}^T g_k = 0$ because of the perfect line search and so $p_k^T g_k = -g_k^T g_k$.

Proposition If $p_0, \ldots, p_k$ are conjugate w.r.t. A then $g_k^T g_{k+1} = 0$.

Proof The definition of p_k implies $g_k = -p_k + \beta p_{k-1}$. Therefore

$$g_k^T g_{k+1} = -p_k^T g_{k+1} + \beta p_{k-1}^T g_{k+1}.$$

But the perfect line search implies $p_k^T g_{k+1} = 0$ and the conjugacy of p_k and p_{k-1} implies $p_{k-1}^T g_{k+1} = 0$. Hence $g_k^T g_{k+1} = 0$.

Proposition If $p_0, \ldots, p_k$ are conjugate w.r.t. A then (11.1.2) makes p_{k+1} conjugate to p_k, that is, $p_k^T A p_{k+1} = 0$.

Proof We know that

$$p_{k+1} = -g_{k+1} + \beta p_k \quad \text{and} \quad A p_k = \frac{1}{s}(g_{k+1} - g_k).$$

Hence

$$p_k^T A p_{k+1} = \frac{1}{s}[-g_{k+1}^T g_{k+1} + g_k^T g_{k+1} + \beta(g_{k+1}^T p_k - g_k^T p_k)].$$

The perfect line searches imply $g_{k+1}^T p_k = 0$; and we have already shown that $g_k^T p_k = -g_k^T g_k$ and $g_k^T g_{k+1} = 0$. Hence

$$p_k^T A p_{k+1} = \frac{1}{s}(-g_{k+1}^T g_{k+1} + \beta g_k^T g_k)$$

which is zero by the definition of β.

Exercise
Complete the steps of a proof by induction which establishes (11.1.4).

11.2.   Conjugate gradients and general functions

The conjugate gradient method can be used to minimize a positive definite quadratic function in at most n iterations from any starting point. We can also modify it as an algorithm for minimizing a general function $F(x)$.

Conjugate gradient method for minimizing $F(x)$

Choose x_0 as an initial estimate of the solution
Calculate $g_0 = \nabla F(x_0)$. Set $p_0 = -g_0$
Repeat for $k = 0, 1, 2, \ldots$
find s by a perfect line search to minimize $F(x_k + sp_k)$
set $x_{k+1} = x_k + sp_k$, $g_k = \nabla F(x_k)$
if k is not a multiple of n then
find β and p_{k+1} from (11.1.2)
else
set $p_{k+1} = -g_{k+1}$
until $\|g_{k+1}\|$ is sufficiently small.

This algorithm proceeds in "cycles" of n iterations, with every n-th search direction being reset as the steepest descent direction. Because we cannot have more than n vectors which are mutually conjugate with respect to a given matrix, each cycle of n steps is regarded as a search for the minimum of a local quadratic model of F. If this does not yield a suitable estimate of the true minimum then a fresh cycle must be started.

The calculation of β in (11.1.2) is called the *Fletcher–Reeves* formula [21]. An alternative, due to Polak and Ribiere [51], is

$$\beta = \frac{g_{k+1}^T(g_{k+1} - g_k)}{g_k^T g_k}. \tag{11.2.1}$$

When F is quadratic (11.1.2) and (11.2.1) give the same β. When F is not quadratic, however, (11.1.2) and (11.2.1) will lead to different search directions. (Of course, when $F(x)$ is not quadratic, the search directions p_k, p_{k-1} are not truly conjugate because there is not a constant Hessian $\nabla^2 F$ for them to be conjugate with respect to.) Other formulae for obtaining conjugate search directions are also given in [26].

Exercises

1. Show that, when applied to a general nonquadratic function, the conjugate gradient method with perfect line searches generates a descent direction on every iteration.

2. Show that the formulae (11.1.2) and (11.2.1) are equivalent when F is a quadratic function.

3. Apply two iterations of the conjugate gradient method to the nonquadratic function $(x_1 - 1)^2 + x_2^3 - x_1 x_2$, starting from the initial point $x = (1, \ 1)^T$.

11.3.   Convergence of conjugate gradient methods

We can establish convergence of the conjugate gradient method using Wolfe's theorem. We can show that p_k is always a descent direction (see Exercise 1 in the previous section) and the perfect line search ensures that (8.1.3), (8.1.4) hold.

In practice the conjugate gradient algorithm usually needs more iterations than a quasi-Newton method. Its ultimate rate of convergence is *n-step quadratic*, which means that

$$||x_k - x^*|| \leq C||x_{k-n} - x^*||^2$$

for some constant C and for k sufficiently large. This implies that convergence will usually be slower than for the Newton and quasi-Newton approaches.

In spite of having slower convergence, the conjugate gradient method does have some potential advantages over Newton and quasi-Newton techniques. Because it does not use any matrices it requires less computer memory when the number of variables, n, is large. Moreover, the number of multiplications per iteration is $O(n)$, compared with $O(n^2)$ for the quasi-Newton method and $O(n^3)$ for the Newton approach. Thus, although it may do more iterations than these matrix-based methods, its overhead cost per iteration may be significantly less.

Convergence of conjugate gradient methods can be accelerated by use of *preconditioning*. Prior to the solution of a system $Ax + b = 0$, transformations can be applied to the matrix A to cause its eigenvalues to become closer together. This is to exploit a stronger finite termination property of the conjugate gradient method which states that the number of iterations required to solve $Ax + b = 0$ will be bounded by the number of distinct eigenvalues of A. For more information on this and on the many other variants of the conjugate gradient approach see [11].

Exercises

1. Estimate the number of multiplications used to evaluate β and calculate a search direction in the conjugate gradient method applied to an n-variable function. Compare this with the number of multiplications used to update the inverse Hessian and calculate a search direction in a quasi-Newton method.

2. The function $F = x_1^2 + x_2^2 + 10x_3^2$ has a Hessian matrix with two equal eigenvalues. Show that the conjugate gradient method converges in two iterations from the starting guess $x_1 = x_2 = x_3 = 1$.

11.4.   Results with conjugate gradients

The OPTIMA implementations of the conjugate-gradient method are denoted by CGp and CGw, signifying, respectively, the use of perfect and weak linesearches. They both use the Fletcher–Reeves recurrence (11.1.2). The theory behind the conjugate-gradient method makes it much more strongly dependent on the use of perfect searches than any of the other minimization techniques we have considered. Indeed there is no theoretical justification for expecting that CGw will converge even in the case when $F(x)$ is quadratic. Tables 11.1 and 11.2 show numbers of iterations and function calls needed by CGp and CGw to solve Problems TD1–OC2. Some points to note from Tables 11.1 and 11.2 are as follows.

• On the quadratic problems VLS1 and OC1(4), CGp behaves like QNp and terminates within n iterations, in agreement with theoretical expectations. When CGw is applied to problems VLS1 and OC1(4), however, its performance is inferior to that of QNw (see Table 9.2). The conjugate gradient approach is more sensitive to the accuracy of the line search.

Problem	Low Accuracy itns/fns	Standard Accuracy itns/fns	High Accuracy itns/fns
TD1	6/34	6/34	7/36
TD2	6/23	6/23	7/25
VLS1	2/5	2/5	2/5
TLS1	6/24	6/24	7/26
VLS2	6/64	6/64	6/64
R1(1)	6/40	7/43	7/43
R1(2)	55/330	63/354	63/354
OC1(4)	3/7	4/10	4/10
OC2(4)	10/29	15/39	18/45

Table 11.1.   CGp solutions for Problems TD1–OC2.

Problem	Low Accuracy itns/fns	Standard Accuracy itns/fns	High Accuracy itns/fns
TD1	12/26	12/26	12/26
TD2	9/18	10/19	14/25
VLS1	2/5	2/5	2/5
TLS1	21/37	24/40	27/44
VLS2	10/40	10/40	12/44
R1(1)	27/79f	27/79f	27/79f
R1(2)	6/9f	6/9f	6/9f
OC1(4)	54/83	78/119	106/161
OC2(4)	15/23	22/34	30/47

Table 11.2.   CGw solutions for Problems TD1–OC2.

- On the nonquadratic problems, conjugate gradient methods are more expensive than the Newton and quasi-Newton approaches. CGw sometimes does better than CGp in terms of function calls but such occasional successes do not justify the use of weak line searches. CGw is much more expensive than CGp on Problems OC1 and OC2 and fails with an uphill search direction on Problems R1(1) and R1(2).
- The n-step quadratic convergence of the conjugate gradient method means that, in practice, it needs more iterations and function calls to go from low- to high-accuracy convergence than does the (superlinearly convergent) quasi-Newton approach. However, the ultimate convergence rates for the conjugate gradient method are better than those for the steepest descent method.

In summary we can say that conjugate gradient methods may have an advantage over Newton or quasi-Newton methods only if their reduced arithmetic cost per iteration can compensate for the extra iterations and function calls they require.

Exercises

1. Use CGp to solve a variant of Problem TD1 in which the target volume is $V^* = 15$.
2. Modify Problem TLS1 to find the best straight-line approximation to the points $(1, 3)$, $(1.1, 3.2)$, $(1.3, 4)$, $(1.6, 4.7)$, $(1.9, 5.7)$ and then find a solution using CGp. Does CGw succeed in solving this problem?
3. Modify Problem VLS2 to find an approximation to the data points
   $$(0, 1), \quad (0.2, 0.95), \quad (0.4, 0.85), \quad (0.6, 0.65), \quad (0.8, 0.35)$$
   using the model $y = \cos(ax + b)$. Attempt this problem using CGp and CGw.
4. Investigate the solutions obtained by CGp applied to Problems R1(1) and R1(2) as ρ increases.
5. Use CGp and CGw to solve Problems OC1(8) and OC2(8). How does their performance compare with that of QNp and QNw?
6. Combine the results from Tables 5.1–11.2 so that for each problem TD1–OC2(4) we can compare the numbers of iterations and function calls needed by all the methods to achieve standard accuracy.
7. Implement a version of the conjugate gradient method which uses the Polak–Ribiere formula (11.2.1) for β rather than the Fletcher–Reeves form. How does it perform on the problems in the first three questions?

11.5. The truncated Newton method

We now describe an approach which combines the Newton and conjugate gradient methods. As explained in Section 9.1, the essential feature of a Newton iteration for minimizing $F(x)$ is the calculation of a search direction, p, from the linear system

$$Gp = -g, \qquad (11.5.1)$$

where $G = \nabla^2 F$ and $g = \nabla F$. However, the solution of (11.5.1) can be computationally expensive and the development of quasi-Newton methods was motivated by the wish to avoid forming and factorizing the exact Hessian.

It can also be argued that we could do less arithmetic and yet retain some benefits of the Newton method if we were to form G as the true Hessian matrix and then obtain p by only approximately solving (11.5.1), using a method significantly cheaper than the Cholesky method. One way of getting such an approximate solution is to apply the conjugate gradient method with a fairly large tolerance on the residuals $\|Gp + g\|$ so that the iteration terminates in appreciably fewer than n steps. The *truncated Newton* approach introduced by Dembo *et al.* [16] makes use of this idea. We give below a version of this algorithm for minimizing a convex function $F(x)$. (This restriction is to ensure that the system $Gp + g = 0$ will always involve a positive definite matrix and hence the conjugate gradient method will be applicable.)

Truncated Newton method for minimizing convex $F(x)$

Choose x_0 as an initial estimate of the solution
Choose C as a constant > 1
Repeat for $k = 0, 1, \ldots$
Calculate $g_k = \nabla F(x_k)$ and $G_k = \nabla^2 F(x_k)$.
Set $\nu_k = \min\{C\|g_k\|, k^{-1}\}$
Apply conjugate gradient iterations to the system $G_k p = -g_k$
and take p_k as the first solution estimate for which $\|G_k p_k + g_k\| < \nu_k$.
Find s so $(x_k + sp_k)$ satisfies Wolfe conditions 2 and 3 for some η_1, η_2
Set $x_{k+1} = x_k + sp_k$
until $\|g_k\|$ is sufficiently small.

The algorithm differs from the standard Newton approach mainly in its use of the parameter ν_k which governs the accuracy with which the Newton system $Gp = -g$ is solved in order to obtain a search direction. The formula for choosing ν_k on each iteration means that it decreases as k increases and as the gradient g_k becomes smaller. Hence p_k tends to the Newton direction as the search gets nearer to an optimum and so the ultimate convergence can be expected to be fast. The potential benefit of the method lies in the fact that it costs less per iteration than the classical Newton technique while the search is still far from a minimum of F.

The truncated Newton method can be extended to apply to a non-convex function $F(x)$. To deal with the possibility that G_k may not be positive-definite on some iterations the inner conjugate gradient iterations must terminate if the calculation of the stepsize, s, from (11.1.3) encounters a denominator that is negative or zero. Additional safeguards may be needed to ensure that any such premature exit from the conjugate gradient solver still yields p_k as a descent direction satisfying Wolfe condition 1.

Exercise

If G is positive-definite, investigate whether the direction p returned after each iteration of a conjugate gradient solution of $Gp = -g$ satisfies the descent condition $p^T g < 0$. What can be said in the case when G is not positive-definite?

Chapter 12

A Summary of Unconstrained Methods

At this, the approximate midpoint of the book, it may be helpful to give
a brief checklist of distinguishing features of the unconstrained optimization methods described so far.

Univariate search
Performs one-dimensional minimizations along each axis in turn.
Can be a direct search or a gradient method.
Convergence is not guaranteed and can be slow.

Hooke and Jeeves method
Adds a pattern move at the end of each cycle of univariate search.
More efficient than basic univariate search.

Nelder and Mead Simplex
Direct search approach.
Explores by moving a "simplex" of trial points in n-dimensional space.
Simplex explores by expansion away from high function values.
Simplex converges by contracting onto a local minimum.
Usually more efficient than univariate search or Hooke and Jeeves method.

DIRECT
Direct search which seeks a global minimum within a hyperbox.
Samples function values at centres of potentially optimal boxes.
Choice of potential optimal sub-boxes is based on Lipschitz constants.
Quite effective at locating global minimum approximately.
Slow convergence if accurate solutions required.

Steepest descent
Uses gradients only.

M. Bartholomew-Biggs, *Nonlinear Optimization with Engineering Applications*,
DOI: 10.1007/978-0-387-78723-7_12, © Springer Science+Business Media, LLC 2008

Works with perfect or weak line search.
Uses $O(n)$ multiplications per iteration.
Ultimate convergence is linear.
Not a very efficient approach.

Newton method
Uses gradient and Hessian matrix.
Works with perfect or weak line search.
Uses $O(n^3)$ multiplications per iteration.
Ultimate convergence is quadratic.
Very efficient approach on convex functions.
May fail (and need backup strategy) when Hessian is not positive-definite.

Gauss–Newton method
Special method for minimizing sums of squared terms.
Uses gradients of individual terms and approximates the Hessian.
Uses $O(n^3)$ multiplications per iteration.
Ultimate convergence can be quadratic in special cases; otherwise it is linear.
Can be more efficient than Newton or quasi-Newton.

Quasi-Newton method
Uses gradient and approximates (inverse) Hessian.
Works with perfect or weak line search.
Uses $O(n^2)$ multiplications per iteration.
Ultimate convergence is superlinear.
Quite effective on convex and nonconvex functions.
Competitive with Newton method when n is large.

Conjugate gradient method
Uses gradients only.
Works with perfect line search only.
Searches along directions which are conjugate w.r.t. the Hessian.
Uses no matrix calculations and takes $O(n)$ multiplications per iteration.
Ultimate convergence is n-step quadratic.
Usually takes more iterations than Newton or quasi-Newton.
Can be efficient in computing effort and memory when n is large.

Chapter 13

Optimization with Restrictions

13.1. Excluding negative variables

We mentioned in Chapter 1 that constraints are often included in
optimization problems, as in (1.1.1) and (1.1.3). However, we have so
far confined ourselves to methods for solving unconstrained problems.
We now show how some relatively simple restrictions on optimization
variables can be incorporated into a problem formulation and still give
rise to an unconstrained optimization calculation to be performed by the
methods described in the preceding chapters. We begin with a reformu-
lation of the tank design problems TD1 and TD2 and then introduce a
new application.

   We have already noted in Problem TD1 that spurious and meaningless
solutions can occur if any of the tank dimensions becomes negative. One
way of preventing this from happening is to introduce a transformation
into the form (1.1.2). Suppose now we let the optimization variables be
y_1, y_2 and then let the tank dimensions be defined by

$$x_i = y_i^2, \quad i = 1, 2.$$

Then the tank dimensions cannot be negative and the objective function
in (1.1.2) becomes

$$S = 2y_1^2 y_2^2 + 2V^* y_2^{-2} + V^* y_1^{-2}. \tag{13.1.1}$$

   Hence we can define a new example *Problem TD1s* which involves minimizing (13.1.1), starting from the initial guess $y_1 = y_2 = \sqrt{2}$. The local
minima are at $y_1^* \approx \pm 1.3077$, $y_2^* \approx \pm 1.8493$ which correspond to the
same physical solution as obtained by minimizing (1.1.2).

M. Bartholomew-Biggs, *Nonlinear Optimization with Engineering Applications*,
DOI: 10.1007/978-0-387-78723-7_13, © Springer Science+Business Media, LLC 2008

We can obtain *Problem TD2s* by a similar transformation of the maximum-volume problem TD2. This replaces the objective function (4.3.3) by

$$V = -\frac{y_1^2 y_2^2 (S^* - 2y_1^2 y_2^2)}{2y_1^2 + y_2^2} \qquad (13.1.2)$$

with $S^* = 35$ and starting from $y_1 = y_2 = \sqrt{2}$. Local minima are given by $y_1^* \approx \pm 1.3068$, $y_2^* \approx \pm 1.8481$ which all correspond to the physical solution of Problem TD2.

The software which can be downloaded along with OPTIMA includes programs for solving problems TD1s and TD2s by the unconstrained optimization methods discussed in previous chapters.

Exercises

1. Obtain expressions for the first and second partial derivatives of the functions (13.1.1) and (13.1.2) with respect to the new variables y_1 and y_2.

2. Obtain contour plots of the functions (13.1.1) and (13.1.2).

3. The squared-variable transformation used in (13.1.1) does not prevent the singularity which occurs when either y_1 or y_2 is zero. However if we define $x_i = 0.1 + y_i^2$ then we effectively put a lower limit on the tank dimensions and the function S is bounded above. Write down the expressions derived from (1.1.2) and (4.3.3) when this change of variables is used and then derive the corresponding expressions for their gradient and Hessian.

Solutions of Problems TD1s and TD2s

It would not be surprising if the modified problems TD1s and TD2s turned out to be harder to solve than the original TD1 and TD2 because the squared-variable transformation increases the nonlinearity of the objective functions. This need not necessarily be the case, however. If we use SDp then Problem TD1 is solved (to standard accuracy) in 24 iterations and 83 function calls while for TD1s the corresponding figures are 22 iterations and 74 function calls. On the other hand SDp solves TD2 in 20 iterations and 56 function calls but needs 21 iterations and 64 function calls on problem TD2s.

A more significant difference in performance occurs if we minimize (1.1.2) and (13.1.1) using the equivalent starting points $x_1 = x_2 = 0.01$ and $y_1 = y_2 = 0.1$. To solve the first problem SDp takes 9 iterations (97 function calls); but for the second SDp needs 22 iterations and 116 function calls. A similar comparison using the same pair of starting points shows that SDp minimizes (1.1.3) in 14 iterations and 36 function calls but takes 24 iterations and 94 function calls to minimize (13.1.2).

A final example which demonstrates the usefulness of the squared-variable transformation involves the starting guess $x_1 = x_2 = 4$ for (1.1.2) and the corresponding initial point $y_1 = y_2 = 2$ for (13.1.1). When SDp is applied to (1.1.2) the first iteration takes a step which makes x_1 and x_2 negative and – as explained in Chapter 1 – the search then continues to reduce the objective function by driving the variables towards $-\infty$. (The program TD1 eventually fails with numerical overflow.) A similar failure does not take place, however, when SDp is applied to (13.1.1) and this function is successfully minimized in 8 iterations.

The exercises below allow the reader to observe how Newton, quasi-Newton and conjugate gradient methods perform on Problems TD1s and TD2s.

Exercises
1. Solve Problem TD1s using SDw, NMw, QNp and CGp and compare the numbers of iterations and function evaluations with those required to solve Problem TD1. Make a similar comparison of solutions to problems TD2s and TD2.
2. Investigate (and discuss) the differences in performance of NMw, QNw and CGp when used to minimize (1.1.3) starting from $x_1 = x_2 = 0.01$ and (13.1.2) starting from $y_1 = y_2 = 0.1$.

13.2.   The preventive maintenance problem

Maintenance plays an important part in reducing the operating costs and increasing the working life of any mechanical system, from a family car to a power station. In this section we consider the optimal scheduling of preventive maintenance (PM), basing our approach on the idea that a system which is regularly maintained can have an *effective age* less than its calendar age.

The cost of operating a system can be expected to increase nonlinearly with time. In the early part of a system's life its cost may be near-linear with fuel and raw materials being used at a steady rate per day. But as the system becomes older it may get less efficient and also begin to incur costs due to the need for repairs or adjustments. If it actually breaks down there will be further costs (such as lost production). Let us suppose that we have determined an expression for a function $H(t)$ which gives the total cost of operating the system up to time t.

In practice, preventive maintenance (PM) is used to lengthen the lifetime of a system (and hence to decrease its average running cost). Under the *effective age model* (Kijima *et al.* [41, 42]) we assume that maintenance makes a system's effective age, y, less than its calendar age, t. This means that the operating costs after a PM will depend on $H(y)$ rather than $H(t)$. Hence, if H is a monotonically increasing function, running costs after a PM will be less than would have been the case if it had not been carried out.

Suppose a system enters service at time $t = 0$ and the first PM occurs at time $t_1 = x_1$. Just before this maintenance, the system's effective age y_1 is the same as its calendar age x_1. Immediately after PM, however, the effective age is reduced to $y_1^+ = b_1 x_1$, where b_1 is some constant $(0 < b_1 < 1)$. Then, during the period until the next PM at time t_2, the effective age of the system is given by $y = b_1 x_1 + x$, $0 < x < x_2 = t_2 - t_1$. In particular, the effective age just before the second PM at time t_2 is $y_2 = b_1 x_1 + x_2$.

Immediately after the second PM, the effective age becomes

$$y_2^+ = y_1^+ + b_2 x_2 = b_1 x_1 + b_2 x_2 = y_2 - (1 - b_2)x_2.$$

That is, the effect of maintenance is to undo some of the aging that has taken place since the first PM. More generally, the effective age immediately after the $(k-1)$-th PM is

$$y_{k-1}^+ = y_{k-1} - (1 - b_{k-1})x_{k-1}. \qquad (13.2.1)$$

We can now say that the operating cost between times t_{k-1} and t_k is given by $H(y_k) - H(y_{k-1}^+)$ rather than by $H(t_k) - H(t_{k-1})$. If $n-1$ is the total number of PMs to be performed in a systems's lifetime (i.e., from time $t = 0$ until its replacement at time $t = t_n$) then its total running cost is

$$H(y_1) + \sum_{k=2}^{n} [H(y_k) - H(y_{k-1}^+)].$$

If c_p is the cost of each PM then the total operating and maintenance cost of the system throughout its life is

$$c_p(n - 1) + \{H(y_1) + \sum_{k=2}^{n} [H(y_k) - H(y_{k-1}^+)]\}.$$

If we also allow for the cost, c_r, of system replacement at time t_n then we can write the mean lifetime cost of the system as

$$C = \frac{c_r + c_p(n-1) + \{H(y_1) + \sum_{k=2}^{n} [H(y_k) - H(y_{k-1}^+)]\}}{t_n}. \qquad (13.2.2)$$

To find an optimal PM schedule we want to find values of $t_1, \ldots, t_n$ which will minimize C. We can express this problem in terms of $x_1, \ldots, x_n$ the intervals between PMs. Clearly

$$t_n = \sum_{k=1}^{n} x_k$$

and the value y_k in the numerator is given by

$$y_k = \left(\sum_{j=1}^{k-1} b_j x_j \right) + x_k.$$

Furthermore, by (13.2.1),

$$y_{k-1}^{+} = y_{k-1} + (1 - b_{k-1})x_{k-1} = \left(\sum_{j=1}^{k-2} b_j x_j \right) + (1 - b_{k-1})x_{k-1}.$$

We assume a cubic polynomial form for the operating cost function

$$H(t) = c_m(t + a_2 t^2 + a_3 t^3).$$

We also assume that the age-reduction factors appearing in (13.2.1) are such that $b_k = b =$ constant, for $k = 1, \ldots, n-1$. (In practice, it may be a nontrivial problem of data analysis to derive values for a_2, a_3 and b which accurately reflect the behaviour of a system.)

We can now define *Problem PM1(n)* in which the cost function C is expressed only in terms of *relative* costs of replacement, maintenance and repair. Therefore we minimize

$$C = \frac{\gamma_r + (n-1) + \gamma_m \{\hat{H}(y_1) + \sum_{k=2}^{n}[\hat{H}(y_k) - \hat{H}(y_{k-1}^{+})]\}}{t_n} \qquad (13.2.3)$$

where

$$\gamma_r = \frac{c_r}{c_p}, \quad \gamma_m = \frac{c_m}{c_p} \quad \text{and} \quad \hat{H}(t) = t + a_2 t^2 + a_3 t^3. \qquad (13.2.4)$$

Solutions of PM problems

In order to minimize (13.2.3) we need to choose a value for n, the number of PMs to be performed. The appropriate number of PMs will depend on γ_r and γ_m. If γ_r is large (because the system has a high replacement cost c_r) then we can expect that it will be efficient to extend the system's working life by performing many PMs. On the other hand, as

γ_r decreases, repeated maintenance has less and less economic advantage compared with replacement. For a particular system, defined by values of $\gamma_r, \gamma_m, a_2, a_3$ and b, we will have to determine the optimum value of n by trial and error.

We can now define an example of Problem PM1(n) which uses the values

$$a_2 = 0.075, \quad a_3 = 0.025 \quad \text{and} \quad b = 0.5. \tag{13.2.5}$$

If we take the unit of time as a year, the coefficients a_2, a_3 imply that – without maintenance – the system running costs increase by 10% after one year and by 50% after two years. The cost data values for our example problem are

$$\gamma_r = 1000, \quad \gamma_m = 100 \tag{13.2.6}$$

which indicate that PM is relatively cheap compared with both system replacement and annual running costs. Program PM1 from the OPTIMA software allows us to solve this problem by a range of optimization methods.

If we solve Problem PM1 with $n = 1$ then we find the optimum operating life of the system when no maintenance is performed. With the data (13.2.5), (13.2.6), the minimum value of (13.2.3) is about 398.6 which is obtained if the system is replaced after about 5.4 years. To see how matters can be improved by preventive maintenance we choose $n = 5$ and and solve PM1(5). At the solution the cost is reduced to about 315 by using maintenance intervals

$$x_1 \approx 1.86, \quad x_2 \approx 1.63, \quad x_3 \approx 1.47, \quad x_4 \approx 1.36, \quad x_5 \approx 1.28.$$

This shows that the system lifetime is extended to about 7.6 years. Table 13.1 shows how (13.2.3) and system lifetime change as n increases.

n	Mean Lifetime Cost	Lifetime (years)
1	398.6	5.4
5	315.0	7.6
10	305.4	8.0
15	302.7	8.2
20	301.7	8.24
25	301.3	8.3

Table 13.1. Optimum PM solutions based on (13.2.5), (13.2.6).

Clearly the beneficial effect of each PM decreases as n increases. The inter-maintenance times become shorter as n gets larger so that $0.6 \geq x_k \geq 0.3$ when $n = 20$ and $0.5 \geq x_k \geq 0.25$ when $n = 25$. In practice we might not want a PM schedule which interrupts normal operation very frequently in pursuit of small savings in cost. In the next section we show how to avoid such schedules.

Excluding small intervals in the PM problem

In order to exclude solutions involving very short inter-PM times, we can use a variation of the squared-variable transformation from the previous section. *Problem PM1s(n)* is expressed in terms of artificial variables $v_1, \ldots, v_n$ such that $x_k = x_{min} + v_k^2$, where x_{min} is the smallest acceptable interval between PMs. It involves the unconstrained minimization of $\tilde{C}(v)$, the cost function (13.2.3) rewritten as a function of $v_1, \ldots, v_n$.

The squared-variable transformation in PM1s is also important in preventing breakdown of solutions to the maintenance scheduling problem. If we try to extend Table 13.1 to the case $n = 30$ then the minimization of (13.2.3) may take a step which makes some of the variables x_k negative. This, in turn, causes the cost function (13.2.3) to be negative. As with the tank design problems, it is then impossible for a minimization algorithm to recover and obtain a sensible solution in which all the inter-PM times are positive.

If we choose $x_{min} = 0.5$ (so that PM cannot occur at less than six-monthly intervals) then solutions of Problem PM1s are given in Table 13.2. For $n \leq 10$ the results are the same as those obtained for Problem PM1; but when $n \geq 15$ the restriction on PM intervals begins to take effect. When $n = 15$ the minimum-cost solution has the last eight PMs equally spaced at six-monthly intervals; however, the optimal cost function is only slightly worse than that given in Table 13.1 when there is no lower limit on the times between maintenance. In the cases $n = 20$ and $n = 25$ the solutions returned by PM1s are markedly worse than those produced by PM1.

n	Mean Lifetime Cost	Lifetime (years)
1	398.6	5.4
5	315.0	7.6
10	305.4	8.0
15	302.8	8.2
20	308.7	10.0
25	334.4	12.5

Table 13.2. Optimum PM solutions with PM intervals > 0.5.

Table 13.3 compares the performances of gradient-based optimization methods when applied to Problems PM1 and PM1s in the case $n = 15$ (using the same data as in Tables 13.1 and 13.2). Both problems are started from the same initial guess with all the PM intervals equal to 1.

	PM1	PM1s
NMw	5/6	12/13
NMp	3/10	8/34
QNw	14/38	23/26
QNp	10/82	20/77
CGp	14/95	31/110

Table 13.3.   Performance of NM, QN and CG on Problems PM1 and PM1s.

Table 13.3 shows that, for all the methods considered, the squared-variable transformation makes PM1s a more difficult problem than PM1. The ranking order between the optimization methods is similar to what we have seen in previous examples.

Exercises

1. Obtain and discuss solutions of PM1 and PM1s when the balance of costs among replacement, repair and maintenance are different. What happens, for example, if replacement is even more expensive so that $\gamma_r = 5000$, $\gamma_m = 100$? What happens if minimal repair is not much more costly than PM so that $\gamma_r = 1000$, $\gamma_m = 10$? What happens if both replacement and repair are relatively less expensive so that $\gamma_r = 100$, $\gamma_m = 10$?

2. What is the optimum maintenance schedule using (13.2.5) and (13.2.6) if the minimum allowable PM interval is given by $x_{min} = 0.1$?

3. Carry out a similar comparison to that in Table 13.3 for the case $n = 10$, when both problems should return the same solution.

Chapter 14

Larger-Scale Problems

14.1.   Control problems with many time steps

Most of the examples considered so far have involved only a few variables. Practical optimization problems often deal with very many unknowns and methods which perform well for small problems may become less efficient as the number of variables increases. We now investigate the behaviour of the methods described in Chapters 7–11 as n becomes larger. The problems we use for our comparison are OC1(n) and OC2(n) from Section 4.3 of Chapter 4. Table 14.1 shows numbers of iterations and function calls needed to solve Problem OC1(n) for various values of n. The figures relate to the high-accuracy convergence test (4.3.2).

	$n = 50$	$n = 100$	$n = 200$
NMw	1/2	1/2	1/2
GNw	1/2	1/2	1/2
QNw	104/105	155/158	236/238
QNp	26/73	29/83	57/163
CGp	26/73	39/113	54/158

Table 14.1.   Performance of NM, GN, QN and CG on Problem OC1(n).

We can make the following observations about Table 14.1.

• Both NM and GN converge on the quadratic problem OC1 in just one iteration. No line search is needed and so NMp and GNp would behave in the same way as NMw and GNw.
• Because OC1(n) is a quadratic problem we would expect QNp to terminate in at most n iterations. In fact we see that convergence occurs in considerably fewer than n steps. This is presumably because

M. Bartholomew-Biggs, *Nonlinear Optimization with Engineering Applications*,
DOI: 10.1007/978-0-387-78723-7_14, © Springer Science+Business Media, LLC 2008

the optimum is rather "flat" and the gradient is near zero in quite a large region round the solution. If we run the problems again with the very high accuracy stopping rule $||g||_2 < 10^{-14}\sqrt{n}$ then QNp takes 42 iterations to solve OC1(50) (and from iteration 26 onwards the function value agrees with the optimum to six significant figures). The very high accuracy solutions to OC1(100) and OC1(200) are found in 58 and 114 QNp iterations, respectively. Thus the n-step finite termination property of QNp is sometimes pessimistic because it relates to the number of iterations (in perfect arithmetic) needed to reduce $||g||$ exactly to zero.

• The behaviour of CGp is quite similar to that of QNp.

• On these problems QNw is less efficient than QNp in terms of both iterations and function calls.

Table 14.2 shows numbers of iterations and function calls needed to solve OC2(n) for various values of n. Bracketed numbers for NM show how many iterations involved a non positive-definite Hessian. As with Table 14.1, the figures relate to the high-accuracy convergence test (4.3.2) with $\epsilon = 10^{-6}$.

	$n = 25$	$n = 50$	$n = 100$
NMw	9(4)/33	18(13)/52	120(115)/139
NMp	9(4)/23	20(15)/32	108(103)/267
GNw	10/15	10/16	13/38
GNp	12/28	10/32	9/44
QNw	85/87	162/163	321/322
QNp	45/132	94/259	193/513
CGp	91/222	190/447	476/1086

Table 14.2.   Performance of NM, QN and CG on Problem OC2(n).

Points to note about Table 14.2 are as follows.

• Problem OC2 is not quadratic and therefore we do not expect the Newton and Gauss–Newton methods to converge in one iteration. Nor do we expect QNp or CGp to converge in fewer than n iterations.

• The Newton methods spend most of their effort using the back-up trust region procedure in Section 9.3 in order to traverse a region where the Hessian is not positive-definite. The Gauss–Newton method, however, has no such difficulties because the approximation to the Hessian based on the Jacobian of subfunctions is positive-definite throughout.

• The numbers of iterations and function calls needed by the quasi-Newton methods seem roughly to double when n doubles. This is not necessarily a general pattern followed for all problems.

• CGp behaves in a similar way to QNp when $n = 25$ and 50. However, the conjugate gradient method is more expensive than might have been expected when $n = 100$. This is possibly a reflection of the fact that CGp is sensitive to rounding errors. The conjugacy and orthogonality properties on which the algorithm depends will not be achieved precisely in finite precision arithmetic. This is particularly noticeable when there are large numbers of variables, because this tends to increase the amount of round-off in key calculations such as the computation of scalar products. Hence conjugate gradient methods can converge more slowly than quasi-Newton methods on nonquadratic problems. Even on quadratic problems, the termination properties predicted by theory are not necessarily observed in practice.

Exercise
Construct a table similar to Table 14.2 which shows the performance of optimization methods on Problem PM1s(n) as n increases from 20 to 50. Make comments on the results similar to those following Table 14.2.

14.2. Overhead costs and runtimes

The problem sizes used in the previous section are an order of magnitude larger than our earlier test problems but they are not what would be regarded as genuinely large problems. It is quite common for engineers and scientists to solve problems involving thousands or tens of thousands of variables. Such problems can be tackled by the kinds of methods described in this book; but in order for the computations to be done efficiently it is important to pay careful attention to implementation issues.

As n increases, it is relevant to compare methods not only on the basis of numbers of iterations and function calls but also in terms of the time they need to find a solution. This will depend on the *overhead costs* of each iteration – i.e., the work done in computing a search direction, updating second derivative information and so on. These matters are, in practice, not simply properties of the algorithm but are also affected by the way in which the algorithm is coded.

One of the most significant implementation issues concerns the efficient handling of large matrix computations, such as the factorization of $\nabla^2 F$ in the Newton method. Hessian matrices in large problems are often *sparse*; that is, they may have as many as 90% of their elements equal to zero. There are then great gains to be made in arithmetic efficiency by use of specialised sparse matrix software. This is able to recognize the presence of zero elements and to avoid such pointless computations as $0 + x = x$ and $0 \times x = 0$. Discussion of the ideas behind sparse matrix operations is outside the scope of this book. However, it is mentioned here as just one among several factors which can greatly improve the performance of an algorithm for large-scale optimization.

Other factors which can affect the computational cost of solving an optimization problem include the way that derivatives are calculated (hand-crafted, with automatic differentiation tools or by finite differences) and the kind of line search that is used (perfect or weak, gradient-based or direct search).

The foregoing discussion implies that runtimes may give a picture of an algorithm's efficiency that is quite different from that suggested by counts of iterations or function evaluations. Table 14.3 shows the computing effort required for the solutions quoted in Table 14.1 for problem OC1(n). The figures are runtimes relative to the time taken by the the Newton method NMw. (Absolute runtimes vary from computer to computer whereas relative runtimes are fairly machine independent.)

	$n = 50$	$n = 100$	$n = 200$
GNw	0.03	0.017	0.008
QNw	0.5	0.4	0.32
QNp	0.37	0.2	0.21
CGp	0.39	0.29	0.2

Table 14.3.   Relative runtimes for GN, QN and CG on Problem OC1(n).

Rather unexpectedly, the figures in Table 14.3 show that all the methods are faster than NMw even in spite of the fact that the Newton method converges in only one iteration. This is partly because the single Newton iteration uses $O(n^3)$ multiplications to compute the search direction whereas the quasi-Newton and conjugate gradient methods only use $O(n^2)$ or $O(n)$ multiplications for each iteration. However the main overhead cost in NMw is the computation of the Hessian by finite differences. OPTIMA does this via the central difference formulae (6.2.8), (6.2.9) and so the gradient has to be calculated at $2n$ points. The cost of the Newton method could be reduced by up to a half by using forward differences (although this might give a less accurate computed Hessian). The NMw runtimes might be reduced even more if analytical second derivatives were employed. Without debating this issue further, we simply underline the point made previously that computational performance of an optimization method can be strongly affected by implementation issues not directly related to the theory behind the underlying algorithm.

The extremely good performance of GNw comes about because the Gauss–Newton method, when applied to a quadratic sum-of-squares function, is able to obtain the exact Hessian without the cost of obtaining second derivatives.

We can use the results in Table 14.3 to make predictions of runtimes for problems involving larger values of n. Suppose that $t^i(\text{M}, n)$ denotes the runtime per iteration of method M when applied to an n-variable problem. If we combine Tables 14.1 and 14.3 we can deduce that

$$t^i(\text{QNw}, 50) \approx \frac{0.5}{104} t^i(\text{NMw}, 50) \quad \text{and} \quad t^i(\text{QNw}, 100) \approx \frac{0.4}{155} t^i(\text{NMw}, 100).$$

If we assume that, for some constants, k_N, k_Q,

$$t^i(\text{NMw}, n) \approx k_N n^3 \quad \text{and} \quad t^i(\text{QNw}, n) \approx k_Q n^2$$

then we can deduce that the relative time-per-iteration satisfies

$$\frac{t^i(\text{QNw}, n)}{t^i(\text{NMw}, n)} \approx \frac{k_Q}{k_N} n^{-1}.$$

In other words, if problem size doubles we expect the time per iteration of QNw to be halved relative to the time per iteration of NMw. The measurements above show that

$$\frac{t^i(\text{QNw}, 100)}{t^i(\text{NMw}, 100)} \frac{t^i(\text{NMw}, 50)}{t^i(\text{QNw}, 50)} \approx \frac{0.4}{155} \times \frac{104}{0.5} \approx 0.54$$

which is in good agreement with expectation. A similar calculation gives

$$\frac{t^i(\text{QNw}, 200)}{t^i(\text{NMw}, 200)} \frac{t^i(\text{NMw}, 100)}{t^i(\text{QNw}, 100)} \approx \frac{0.32}{236} \times \frac{155}{0.4} \approx 0.53$$

and so the theory still holds quite well. Similar analysis can be done with regard to the times per iteration for QNp and CGp (see exercises below).

We note that CGp is slightly faster than QNp when $n = 200$. Because the overhead costs for CGp are $O(n)$ and those for QNp are $O(n^2)$ we can conjecture that CGp may be significantly faster than QNp as n gets larger, even if the conjugate gradient method requires more iterations. (This conjecture can be tested by running the OPTIMA software.)

Table 14.4 shows runtimes for solving Problem OC2(n) relative to the time taken by NMw. Perhaps the most striking feature is that GN, QN and CG are all much faster than NM for this nonquadratic problem. The cost of forming and factorizing $\nabla^2 F$ is now incurred many times because the Newton method takes more than one iteration to converge. Clearly NMp and NMw could do better if the Hessian were handled more efficiently. In other words, the OPTIMA implementation of the Newton method could almost certainly be improved. The results presented here

	$n = 25$	$n = 50$	$n = 100$
NMp	1.04	0.93	0.93
GNw	0.075	0.01	0.002
GNp	0.1	0.021	0.0023
QNw	0.225	0.082	0.014
QNp	0.29	0.13	0.021
CGp	0.46	0.21	0.043

Table 14.4.   Relative runtimes for GN, QN and CG on Problem OC2(n).

show that implementation issues can be as important as the theoretical properties of an algorithm.

The pattern of behaviour among the other methods is broadly similar to that in Table 14.3 except that QNp is now slower than QNw. This is due to the relative numbers of iterations. On the quadratic problem OC1, QNw uses four or five times as many iterations as QNp whereas on Problem OC2, QNw takes only about twice as many. The savings made by QNp on iteration count are now not sufficient to outweigh the arithmetic costs of the perfect line search.

Exercises

1. Run NMp on Problems OC1(50), OC1(100), OC1(200) and add the results to the comparisons in this section.
2. Using the data in Tables 14.1 and 14.3, consider the timings $t^i(\text{CGp}, n)$ in a similar way to those discussed in the main text for QNw. Hence deduce a value for the ratios

$$\frac{t^i(\text{CGp}, n)}{t^i(\text{QNw}, n)} \quad \text{and} \quad \frac{t^i(\text{CGp}, n)}{t^i(\text{QNp}, n)}.$$

3. Repeat the analysis of Question 2 using results from Tables 14.2 and 14.4.
4. Run program OC1 to see if CGp runs faster than QNp when applied to Problem OC1(n) with $n > 200$.
5. Consider a modified version of problem OC2(n) which uses the data

$$t_f = 5, \quad \tau = \frac{t_f}{n}, \quad u_0 = s_0 = 0, \quad s_f = 4, \quad u_f = 1$$

and perform a comparison similar to that in Tables 14.2 and 14.4 by solving it for $n = 50, 75, 100$.

6. Perform a comparison similar to that in Tables 14.2 and 14.4 based on solving Problems PM1(n) and PM1s(n) for $n = 20, 30, 40$.

Chapter 15

Global Unconstrained Optimization

A practical optimization problem may have several local solutions. We have already seen this in the case of the routing problem R1(1). Contours of the objective function for R1(1) are illustrated in Figure 4.9 and show that there are two locally optimal routes which pass on different sides of the obstacle. As discussed in Section 4.3 of Chapter 4, one of these local solutions gives a lower objective function value than the other and would be regarded as the global minimum.

For problems in more than two variables we cannot plot contours and hence it is not easy to detect multiple solutions. Therefore, when we apply one of the minimization methods described in previous chapters, we cannot usually be certain whether it has terminated at a local or a global solution. Unfortunately there are no computable conditions which will, in general, establish whether x^* is a global optimum. Hence the global optimization problem is inherently more difficult than the problem of finding any local minimum.

Methods for tackling the global optimization problem cannot in general be guaranteed to be successful. In practice they will usually terminate at a point which has a fairly high degree of probability of being the global solution and they may require a considerable amount of computing effort even to achieve this much.

We have already introduced one global optimization technique: DIRECT, described in Chapter 5. This uses function values only and hence is suitable for nonsmooth problems. In the next section we consider an alternative approach which can use gradient information. For a much fuller account of the global optimization problem and its solution see [26].

M. Bartholomew-Biggs, *Nonlinear Optimization with Engineering Applications*,
DOI: 10.1007/978-0-387-78723-7_15, © Springer Science+Business Media, LLC 2008

15.1. Multistart methods

A heuristic approach to the global minimization problem would be to run a local minimization algorithm from many different starting points and then pick the best of the solutions. This strategy can sometimes be effective, but its drawbacks are *(i)* it is wastefully expensive because many local searches may yield the same result; and *(ii)* it provides no assurance that the local optima found do actually include the global solution. The approach can be formalised and made more efficient by the incorporation of some statistical theory. To illustrate this we mention two ideas which are used in a global optimization method proposed by Rinnooy-Kan and Timmer [57, 58].

Cluster analysis can be used to see if different local optimizations are tending to the same result. If we allow all the optimizations to perform a fixed number of iterations (not too large) we can estimate how many of the searches seem to be heading for different solutions. We can then perform another set of iterations of a (probably much smaller) number of optimization calculations and repeat the cluster analysis. Continuing in this way, we would expect to locate multiple minima more cheaply than with the basic "scattergun" approach.

Bayesian estimation can be used to determine an expected number of minima of the objective function on the basis of the number found so far (W) and the number of local searches used (N_s). A formula for the expected total number of local solutions is

$$W_t = \frac{W(N_s - 1)}{N_s - W - 2}.$$
(15.1.1)

Thus, if 5 minima are found in 30 searches, we get $W_t \approx 6.3$, which suggests that further solutions may exist. If no more minima are found when 100 searches have been completed then $W_t \approx 5.3$ and it is now more reasonable to suppose that there are only five local solutions.

The global optimization algorithm given in [57, 58] uses both these ideas. An initial iteration (using N_s starting points and clustering) produces W local optima, say. If $W_t \gg W$ then further cycles of local optimization and clustering are performed from new starting points until $W_t < W + 0.5$ (say). The algorithm also includes strategies, not described here, to ensure that additional starting points are not chosen too close to minima that have already been found or to starting points that have been used previously.

15.2.   Global solution of routing problems

Consider Problem R1(1) (see Section 4.3) and suppose we seek the optimal turning point in the box with corners (0,0), (8,8). We first adopt a very simplified multistart approach and perform a quasi-Newton minimization from twenty randomly chosen starting points in this search region.

Using the program R1g from the OPTIMA software we find the local solution $x^* \approx 3.03$, $y^* \approx 5.31$, $F^* \approx 11.27$ on four occasions and the global minimum $x^* \approx 4.98$, $y^* \approx 1.18$, $F^* \approx 9.25$ on the remaining sixteen trials. Using the formula (15.1.1) it follows that the expected total number of local minima is

$$W_t = \frac{2 \times 19}{16} \approx 2.3.$$

Hence, even if we did not already know that there were only two minima, it would be reasonable to conclude that there were no further solutions. The twenty minimizations use a total of 232 quasi-Newton iterations and 474 function and gradient calls. We can apply DIRECT to the same problem and, with the parameter $\epsilon = 0.01$, it converges in 43 iterations and 421 function calls to

$$x^* \approx 4.95, \quad y^* \approx 1.17, \quad F^* \approx 9.25$$

which is close to the global optimum. We do not normally expect DIRECT to produce high-accuracy solutions (because it does not use gradient information and only samples the objective function at discrete points). We can usually improve on the best point returned by DIRECT if we use it as a starting point for a local quasi-Newton search. In this case, the local search gives accurate values for x^* and y^* in just 6 iterations and 12 function and gradient calls.

We now turn to *Problem R2g* involving two circular obstacles and two turning points. The route is from (0,0) to (10, 4.5) and the first obstacle is centred on (4,3) with radius 2 while the second has centre (8,4) and radius 1. The search region for the first turning point is the box with corners (0,0) and (6 ,8) and the search region for the second turning point has corners (6,0) and (12, 8). There are three local solutions, shown in Figure 15.1. The route OABP is the global solution where the route-cost function ≈ 11.17. Route OCDP has a cost about 12.97 and the over-and-under route OEFP costs about 14.99.

Using the OPTIMA program R2g, twenty random starts of the quasi-Newton method yield route OCDP ten times. The global optimum is found eight times and the worst route OEFP only once. Interestingly,

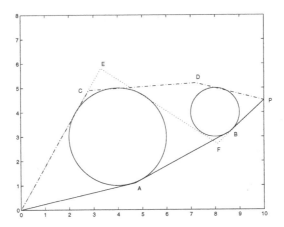

Figure 15.1.   Multiple solutions of a two-obstacle routing problem.

one of the quasi-Newton runs terminates with the first turning point almost at (0,0) and the second at (6.94, 1.64). This solution gives a cost function value of about 11.32 which is the optimal route with only one turning point (because the first turn has effectively been eliminated). On the evidence of these twenty local searches the formula (15.1.1) suggests that there are 5.4 minima in total. In order to be reasonably certain that there are no further solutions we would need to perform over thirty further quasi-Newton minimizations.

The total cost of the twenty randomly-started minimizations is 747 quasi-Newton iterations and 2618 function and gradient calls. Once again, DIRECT appears more efficient because it takes 126 iterations and 997 function calls to find a good estimate of route OABP (with cost function value $\approx$ 11.23). Using this as a starting point, we can find the gobal minimum accurately by one further application of the quasi-Newton method (using 17 iterations and 103 function and gradient calls).

Exercise
Solve variants of Problem R2g to investigate how the optimum route changes as the target point changes in the range between (10, 2) and (10, 7).

15.3. Global solution of a feed-blending problem

Suppose that a brand of animal feed is to be produced from n ingredients. Let $c_1, \ldots, c_n$ denote the costs per kilogram of each ingredient. Suppose also that the feed must meet a specification for nutrient content (i.e., levels of vitamins, fat, fibre, and so on). We assume that 1 kg of ingredient i contains α_{ji} kg of nutrient j (where $1 \le j \le m$). If the feed is to be produced in 50 kg bags then the amount, x_i, of ingredient i to be packed in each bag must satisfy

$$\sum_{i=1}^{n} x_i = 50; \quad \text{and} \quad \sum_{i=1}^{n} \alpha_{ji} x_i = \beta_j \quad \text{for} \quad j = 1, \ldots, m \quad (15.3.1)$$

where the values β_j are given nutrient specifications.

If $m < n$, the conditions (15.3.1) do not determine the x_i uniquely. Therefore we can seek the x_i to minimize the cost of producing a feed which meets the nutrient specifications as closely as possible. If we minimize the function

$$C(x) = \sum_{i=1}^{n} c_i x_i + \rho \left[\left(\sum_{i=1}^{n} x_i - 50 \right)^2 + \sum_{j=1}^{m} \left(\sum_{i=1}^{n} \alpha_{ji} x_i - \beta_j \right)^2 \right],$$

$$(15.3.2)$$

for some positive weighting factor ρ, then we get a low-cost mixture which takes account of nutrient requirements. By increasing ρ we enforce more strongly the satisfaction of these requirements.

$C(x)$ in (15.3.2) is a quadratic function of the x_i and so is easy to minimize. However, as with the tank design and preventive maintenance problems, there is a possibility that some of the x_i will be negative at the minimum of (15.3.2). To avoid such impractical solutions we could use the squared-variable transformation introduced in Chapter 13. Thus would involve minimizing the non-quadratic function

$$\tilde{C}(y) = \sum_{i=1}^{n} c_i y_i^2 + \rho \left[\left(\sum_{i=1}^{n} y_i^2 - 50 \right)^2 + \sum_{j=1}^{m} \left(\sum_{i=1}^{n} \alpha_{ji} y_i^2 - \beta_j \right)^2 \right]$$

$$(15.3.3)$$

and then setting $x_i = y_i^2$, $i = 1, \ldots, n$.

The function (15.3.3) will have multiple minima because for any local solution defined by the values $\hat{y}_1, \ldots, \hat{y}_n$ there will also be a solution at $-\hat{y}_1, \ldots, \hat{y}_n$. (The same is true for sign changes in any or all of the variables.) However, all such local minima are equivalent in that they yield the same value of $\tilde{C}$. We now consider a more interesting version of the feed-blending problem in which there may be multiple minima with different objective function values.

Suppose that, for convenience in the production process, we do not want the feed to contain very small amounts of any ingredient which might be difficult, in practice, to measure accurately. Therefore we would like to look for a mix of ingredients which meets the nutrient specification at low cost while also satisfying a restriction of the form

$$\text{either} \quad x_i = 0 \quad \text{or} \quad x_i \geq x_{min}.$$

We can attempt to solve this feed-blending problem by minimizing an extended form of (15.3.2), namely

$$\hat{C}(x) = C(x) + \hat{\rho} \sum_{i=1}^{n} \psi(x_i)^2 \qquad (15.3.4)$$

where

$$\psi(x_i) = \begin{cases} 0 & \text{if } x_i > x_{min} \\ 4x_i(x_{min} - x_i)/x_{min}^2 & \text{if } 0 \leq x_i \leq x_{min}. \end{cases} \qquad (15.3.5)$$

The function $\psi(x_i)$ takes values between zero and one and is used to penalise any x_i values which lie in the unacceptable range between 0 and x_{min}. The function (15.3.4) is likely to have several local minima – each corresponding to some x_i being close to zero or x_{min} – and so we need to approach it using a global minimization technique. *Problem FBg* involves (15.3.4) and (15.3.5) in the case when $n = 3$ and $m = 1$. The cost function coefficients are such that (15.3.2) is

$$C(x) = 1.5x_1 + x_2 + 0.8x_3 + 80(x_1 + x_2 + x_3 - 50)^2$$

$$+ \; 80(0.12x_1 + 0.08x_2 + 0.06x_3 - 3.75)^2.$$

The function $\hat{C}$ to be minimized is then given by (15.3.4), (15.3.5) with $\hat{\rho} = 4$ and $x_{min} = 1$.

If we solve FBg using both simplified multistart and DIRECT the results are as follows. Twenty quasi-Newton minimizations from random starting points in the range $0 \leq x_i \leq 50$ give three candidate local minima:

$$(i) \quad x_1 \approx 0, \quad x_2 \approx 34.4, \quad x_3 \approx 15.6; \quad \hat{C} \approx 47.2$$

$$(ii) \quad x_1 \approx 1, \quad x_2 \approx 31.4, \quad x_3 \approx 17.6; \quad \hat{C} \approx 47.3$$

$$(iii) \quad x_1 \approx 11.3, \quad x_2 \approx 0, \quad x_3 \approx 38.7; \quad \hat{C} \approx 48.3$$

The second solution is found eighteen times out of the twenty trials and the others occur only once each.

When DIRECT is applied to the problem, starting from the midpoint of the hyperbox $0 \leq x_i \leq 50$, it terminates at a point close to local minimum *(ii)* and a subsequent quasi-Newton refinement locates solution *(ii)* exactly. This example illustrates the fact that DIRECT may not do so well on problems where the global minimum has a function value which is quite close to the function value at one or more of the local solutions. The exploration technique used by DIRECT will not easily identify potentially optimal regions which offer only a small improvement to the currently best solution estimate.

Exercise

Use FBg to do a comparison between DIRECT and the quasi-Newton multistart approach when the range for the variables is $0 \le x_i \le 40$.

15.4. Global solution of a sensitivity problem

When a function $F(x)$ has been minimized we might wish to know how much the optimal values of the variables could be changed without causing a more than 1% increase in the function value. This could be important if the variables are physical dimensions and we need to set manufacturing tolerances.

Consider Problem TD1 whose solution is $x_1^* = 1.71, x_2^* = 3.42$, giving a minimum surface area $S(x^*) = 35.09$ (given by (1.1.2) with $V^* = 20$). To estimate the smallest change to the variables that will cause the surface area to increase by 1% we can minimize

$$F(x) = (x_1 - x_1^*)^2 + (x_2 - x_2^*)^2 + \rho(2x_1x_2 + 40x_2^{-1} + 20x_1^{-1} - 35.44)^2.$$
$$(15.4.1)$$

Figure 15.2 shows the contours of this function. There are two local minima near the points marked A and B. There is a local maximum near the point C; and in the regions around D and E there are saddle points.

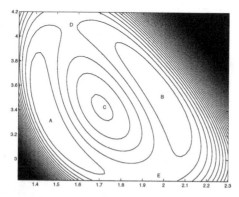

Figure 15.2. Multiple solutions of a sensitivity problem for TD1.

The left-hand contour plot in Figure 4.4 shows why there are two local minima. The contour lines around the solution of Problem TD1 indicate that the objective function increases most rapidly in roughly the east-west direction. The contour lines are not symmetrical about the minimum but are flattened more on one side than the other. Hence any small step away from x^* in the eastward direction will produce a bigger increase in the objective function than a step of the same size in the westward direction. In particular, for a specified increase in the function there will be a minimum length step in the eastward direction and another, different, minimum length step in a westward direction.

Exercises

1. Use any minimization method together with a range of starting points to find all the local minima of (15.4.1).

2. Construct a contour plot similar to Figure 15.2 for the function whose minimum estimates the maximum change to the solution of Problem TD1 which will produce a 1% increase in surface area.

Chapter 16

Equality Constrained Optimization

16.1.   Problems with equality constraints

So far, we have dealt only with methods for solving unconstrained optimization problems. However, as shown in Chapter 1, we can also express the minimum surface area problem TD1 in the form *Problem TD1a*

$$\text{Minimize} \quad 2x_1x_2 + 2x_1x_3 + x_2x_3 \quad \text{subject to} \quad x_1x_2x_3 = V^* \quad (16.1.1)$$

which involves an objective function and a nonlinear equality constraint. Similarly the maximum volume problem TD2 can be written as *Problem TD2a*

$$\text{Minimize} \quad -x_1x_2x_3 \quad \text{subject to} \quad 2x_1x_2 + 2x_1x_3 + x_2x_3 = S^*. \quad (16.1.2)$$

(Both TD1a and TD2a could also be formulated using the $x_i = y_i^2$ transformation from Chapter 13 to prevent negative dimensions occurring. However there are better ways of dealing with this issue using inequality constraints. These are considered in a later chapter.)

Other equality constrained problems can be based on the least squares examples introduced in Chapter 3. If we wish to force the model curve to pass through one or more of the data points we can use modified forms of VLS1 and VLS2, such as

Problem VLS1a

$$\text{Minimize} \quad \sum_{i=2}^{m} (z_i - x_1 - x_2 t_i)^2 \quad \text{subject to} \quad z_1 - x_1 - x_2 t_1 = 0. \quad (16.1.3)$$

M. Bartholomew-Biggs, *Nonlinear Optimization with Engineering Applications*,
DOI: 10.1007/978-0-387-78723-7_16, © Springer Science+Business Media, LLC 2008

Problem VLS2a

$$\text{Minimize} \quad \sum_{i=1}^{m-1}(z_i - x_1 e^{x_2 t_i})^2 \quad \text{subject to} \quad z_m - x_1 e^{x_2 t_m} = 0. \quad (16.1.4)$$

We can treat also total least squares approximation as a constrained optimization problem. Suppose we have data points $(t_1, \ z_1), \ldots,$ $(t_m, \ z_m)$ and a model function $z = \phi(x, t)$. Then, as pointed out in Section 3.1, the *footpoint*, t_f, corresponding to the ith data point solves the unconstrained problem

$$\text{Minimize} \quad \Psi(t_f) = (t_i - t_f)^2 + (z_i - \phi(x, t_f))^2.$$

Hence t_f must satisfy

$$\frac{d\Psi}{dt_f} = 0$$

which leads to

$$(t_i - t_f) + (z_i - \phi(x, t_f))\phi_t(x, t_f) = 0 \qquad (16.1.5)$$

where ϕ_t denotes the first partial derivative of ϕ w.r.t. t. In forming the problem TLS1 we obtained t_f by solving this equation. We could, however, regard (16.1.5) as a constraint and treat each footpoint as an additional variable. If we let τ_i denote the footpoint for the ith data point then, by adapting (3.1.5) and (3.1.8), we can solve the total least squares problem by treating $x_1, \ldots, x_n$ and $\tau_1, \ldots, \tau_m$ as variables and minimizing

$$\sum_{i=1}^{m}(t_i - \tau_i)^2 + (z_i - \phi(x, \tau_i))^2$$

subject to $(t_i - \tau_i) + (z_i - \phi(x, \tau_i))\phi_t(x, \tau_i) = 0 \quad$ for $\quad i = 1, \ldots, m.$

If $\phi(x, t)$ is the linear model $z = x_1 + x_2 t$ then, on letting $x_{i+2} = \tau_i$, we obtain *Problem TLS1a*

$$\text{Minimize} \quad \sum_{i=1}^{m}(t_i - x_{i+2})^2 + (z_i - x_1 - x_2 x_{i+2})^2 \qquad (16.1.6)$$

subject to $(t_i - x_{i+2}) + (z_i - x_1 - x_2 x_{i+2})x_2 = 0 \quad$ for $\quad i = 1, \ldots, m.$
$$\qquad (16.1.7)$$

In a similar way, the exponential model from VLS2 leads to *Problem TLS2a*

$$\text{Minimize} \quad \sum_{i=1}^{m}(t_i - x_{i+2})^2 + (z_i - x_1 e^{x_2 x_{i+2}})^2 \qquad (16.1.8)$$

subject to

$$(t_i - x_{i+2}) + (z_i - x_1 e^{x_2 x_{i+2}}) x_1 x_2 e^{x_2 x_{i+2}} = 0 \quad \text{for} \ \ i = 1, \dots, m.$$

$$(16.1.9)$$

We can consider variants of the optimal control problems from Section 3.3, in which the terminal conditions are treated explicitly as constraints rather than simply being included with the smoothness conditions in a weighted sum-of-squares function. *Problem OC1a(n)* is

$$\text{Minimize} \quad x_1^2 + x_n^2 + \sum_{k=2}^{n} (x_k - x_{k-1})^2 \quad \text{s.t.} \quad s_n = s_f \ \text{and} \ u_n = u_f$$

$$(16.1.10)$$

where s_n and u_n are given by (3.3.1). *Problem OC2a(n)* is of the same form as (16.1.10) but has a different objective function, namely

$$\text{Minimize} \quad x_1^2 + x_n^2 + \sum_{k=2}^{n} \left(1 - \frac{x_k}{x_{k-1}}\right)^2 \quad \text{s.t.} \quad s_n = s_f \ \text{and} \ u_n = u_f.$$

$$(16.1.11)$$

We can also consider an optimal control problem which includes a drag term. We make the fairly common assumption that drag is proportional to speed squared and so suppose that the actual acceleration during the k-th timestep is modelled by $x_k - c_D u_k^2$ where x_k is the applied acceleration and c_D is a drag coefficient. *Problem OC3(n)* has $x_1, \dots, x_n$ and $u_1, \dots, u_{n-1}$ as variables and the objective function is, as before,

$$F = x_1^2 + x_n^2 + \sum_{i=2}^{n} (x_i - x_{i-1})^2. \qquad (16.1.12)$$

There are now $n + 1$ equality constraints

$$u_k - u_{k-1} - (x_k - c_D u_k^2)\tau = 0 \quad \text{for} \ \ k = 1, \dots, n - 1 \quad (16.1.13)$$

$$u_f - u_{n-1} - (x_n - c_D u_f^2)\tau = 0 \qquad (16.1.14)$$

and

$$s_n - s_f = 0 \qquad (16.1.15)$$

where s_n is given by the recurrence relation

$$s_k = s_{k-1} + u_{k-1}\tau + \frac{1}{2}(x_k - c_D u_k^2)\tau^2 \quad \text{for} \ \ k = 1, \dots, n. \quad (16.1.16)$$

Problem FBc is a version of the feed-blending problem from Chapter 15. It involves the minimization of

$$F(x) = \sum_{i=1}^{n} c_i x_i + \sum_{i=1}^{n} \psi(x_i)^2 \qquad (16.1.17)$$

where $\psi(x_i)$ is given by (15.3.5). The constraints are

$$\sum_{i=1}^{n} x_i = 50 \quad \text{and} \quad \sum_{i=1}^{n} \alpha_{ji} x_i = \beta_j \quad \text{for } j = 1, \ldots, m. \qquad (16.1.18)$$

The routing problem from Chapter 3 can be rewritten using constraints to enforce the condition that the vehicle must not enter the no-go regions. Using the notation of Section 3.2, *Problem R1(2)c* is

Minimize $d(0, 0, x_1, y_1) + d(x_1, y_1, x_2, y_2) + d(x_2, y_2, x_3, y_3)$ (16.1.19)

subject to

$$\nu(0, 0, x_1, y_1) = 0; \quad \nu(x_1, y_1, x_2, y_2) = 0; \quad \nu(x_2, y_2, x_3, y_3) = 0. \qquad (16.1.20)$$

As we show in Section 16.5, however, this problem presents some difficulties which are not present in the other examples of this section.

16.2.   Optimality conditions

All the problems in the previous section are instances of the general equality-constrained minimization or *nonlinear programming* problem, namely

$$\text{Minimize } F(x) \qquad (16.2.1)$$

$$\text{subject to } c_i(x) = 0 \quad i = 1, \ldots, l. \qquad (16.2.2)$$

Definition If x satisfies the constraints (16.2.2) it is said to be *feasible*. Otherwise it is called *infeasible*.

If x^* is a solution of (16.2.1), (16.2.2) then it must be a feasible point. The optimality of x^* can be thought of as a balance between the function and the constraints. By this we mean that a move away from x^* cannot be made without either violating a constraint or increasing the function value. This can be stated formally as follows.

Proposition If x^* solves (16.2.1), (16.2.2) and $x^* + \delta x$ is a nearby point then
    (i) if $F(x^* + \delta x) < F(x^*)$ then $c_i(x^* + \delta x) \neq 0$ for some i
    (ii) if $c_1(x^* + \delta x) = \cdots = c_l(x^* + \delta x) = 0$ then $F(x^* + \delta x) \geq F(x^*)$.

First-order conditions

The following proposition (stated without proof) gives optimality conditions for an equality constrained problem when the function and constraints are differentiable. These are called the *Karush–Kuhn–Tucker* (KKT) conditions.

Proposition If x^* is a local solution of (16.2.1), (16.2.2) then *(i)* the point x^* must be feasible and so

$$c_i(x^*) = 0 \quad (i = 1, \ldots, l) \tag{16.2.3}$$

and *(ii)* there must exist scalars $\lambda_1^*, \ldots, \lambda_l^*$ such that

$$\nabla F(x^*) - \sum_{i=1}^{l} \lambda_i^* \nabla c_i(x^*) = 0. \tag{16.2.4}$$

Definition The scalars $\lambda_1^*, \ldots, \lambda_l^*$ in (16.2.4) are called *Lagrange multipliers*.

Definition The vectors $\nabla c_1(x), \ldots, \nabla c_l(x)$ are called the *constraint normals*.

For the Lagrange multipliers λ_i^* to be unique the constraint normals $\nabla c_i(x^*)$ must be linearly independent. (See Exercise 4 at the end of this section and Exercise 2 in Section 16.3.)

Definition The $l \times n$ matrix with rows $\nabla c_1(x)^T, \ldots, \nabla c_l(x)^T$ is known as the *Jacobian* of the constraints.

If N is the Jacobian of the constraints (16.2.2) then (16.2.4) can be written

$$\nabla F(x^*) - N^T \lambda^* = 0. \tag{16.2.5}$$

Conditions (16.2.4) and (16.2.5) imply that $\nabla F(x^*)$ is linearly dependent on the constraint normals. This reflects the fact that a constrained minimum occurs when the gradients of the function and the constraints interact in such a way that any reduction in F can only be obtained by violating the constraints.

The left-hand side of (16.2.4) can be regarded as the gradient of a function

$$L(x, \lambda^*) = F(x) - \sum_{i=1}^{l} \lambda_i^* \, c_i(x) = F(x) - \lambda^{*T} c(x). \tag{16.2.6}$$

Definition $L(x, \lambda^*)$ is the *Lagrangian* function for problem (16.2.1), (16.2.2).

Feasible directions and second-order conditions

Definition An n-vector, z, is said to be a *feasible direction* at x^* if $Nz = 0$, where N is the matrix of constraint normals appearing in (16.2.5).

Let us assume z is a feasible direction normalized so that $||z|| = 1$. If we consider the Taylor expansion

$$c(x^* + \epsilon z) = c(x^*) + \epsilon Nz + O(||\epsilon z||^2)$$

then $c(x^* + \epsilon z) = O(\epsilon^2)$. Therefore a move away from x^* along z keeps the constraints satisfied to first-order accuracy. In particular, if all the constraints (16.2.2) are linear then $x^* + \epsilon z$ is a feasible point for all ϵ. If any of the $c_i(x)$ are nonlinear then z defines a direction *tangential* to the constraints at x^*.

Proposition Condition (16.2.5) implies that, for any feasible direction z,

$$z^T \nabla F(x^*) = 0. \tag{16.2.7}$$

Proof The result follows on premultiplying (16.2.5) by z^T.

Expressions (16.2.3) and (16.2.4) are first-order conditions that hold at any constrained stationary point. To distinguish a minimum from a maximum or a saddle point we need a second-order condition which can be stated as follows.

Proposition When the constraint functions c_i are all linear, the second-order condition guaranteeing that x^* is a minimum of problem (16.2.1), (16.2.2) is

$$z^T \nabla^2 F(x^*) z > 0 \tag{16.2.8}$$

for any feasible direction z.

For problems with nonlinear constraints it is the Hessian of the Lagrangian function (16.2.6) which appears in the second order optimality condition.

Proposition When the constraint functions are nonlinear, the second order condition that guarantees x^* is a minimum of problem (16.2.1), (16.2.2) is

$$z^T \nabla^2 L(x^*, \lambda^*) z > 0 \tag{16.2.9}$$

for any feasible direction z.

Exercises

1. Use Taylor series arguments to show that for a problem with all linear constraints the optimality conditions (16.2.3), (16.2.4) and (16.2.8) ensure that if δx is such that if $c_1(x^* + \delta x) = \cdots = c_l(x^* + \delta x) = 0$ then

$$F(x^* + \delta x) \geq F(x^*). \qquad (16.2.10)$$

If the constraints are all quadratic show that conditions (16.2.3), (16.2.4) and (16.2.9) will cause (16.2.10) to hold when $c_i(x^* + \delta x) = \cdots = c_l(x^* + \delta x) = 0$.

2. Show that, if G is positive definite, the problem

$$\text{Minimize } \frac{1}{2}x^T G x + h^T x \quad \text{subject to } x^T x = 1$$

has a solution given by $x = -(\lambda I + G)^{-1}h$ for some scalar λ. How does this result relate to trust region methods (Chapter 9)?

3. Show that, if the constraints (16.2.2) are all divided by a constant factor k, the solution of the modified nonlinear programming problem is unchanged except that the new Lagrange multipliers are given by $k\lambda_1^*, \ldots, k\lambda_l^*$.

4. If (16.2.1), (16.2.2) has a solution x^* where the constraint normals are not linearly independent show that $\lambda_1^*, \ldots, \lambda_l^*$ are not uniquely defined.

16.3. A worked example

In some cases the optimality conditions (16.2.3), (16.2.4) can be used directly to find a solution (x^*, λ^*). Consider the problem

$$\text{Minimize } F(x) = x_1^2 + 3x_1x_2 \quad \text{subject to } c_1(x) = x_1 + 5x_2 - 1 = 0.$$
$$(16.3.1)$$

The optimality conditions mean that x_1^*, x_2^* and λ_1^* satisfy the three equations

$$x_1 + 5x_2 - 1 = 0$$

$$\frac{\partial F}{\partial x_1} - \lambda_1 \frac{\partial c_1}{\partial x_1} = 2x_1 + 3x_2 - \lambda_1 = 0$$

$$\frac{\partial F}{\partial x_2} - \lambda_1 \frac{\partial c_1}{\partial x_2} = 3x_1 - 5\lambda_1 = 0.$$

From the last equation we get $\lambda_1 = 3x_1/5$ and then the second equation gives $x_2 = -7x_1/15$. Hence the first equation reduces to $-4x_1/3 - 1 = 0$

and so the constrained minimum occurs at

$$x_1^* = -\frac{3}{4}, \quad x_2^* = \frac{7}{20} \quad \text{with Lagrange multiplier} \quad \lambda_1^* = -\frac{9}{20}.$$

We can confirm that x^* satisfies the second-order optimality conditions if we can show that $z^T G z > 0$ when z is a feasible direction. The constraint normal matrix is $N = (1,\ 5)$ and so $z = (-1,\ 0.2)^T$ is a feasible direction because

$$N z = (1,\ 5) \begin{pmatrix} -1 \\ 0.2 \end{pmatrix} = 0.$$

In fact z is the *only* feasible direction and

$$z^T G z = (-1\ \ 0.2) \begin{pmatrix} 2 & 3 \\ 3 & 0 \end{pmatrix} \begin{pmatrix} -1 \\ 0.2 \end{pmatrix} = 0.8.$$

Because this is positive we can be sure that we have found a constrained minimum (rather than a maximum or saddle point).

Exercises

1. Use the optimality conditions to solve

$$\text{Minimize} \quad x_1^2 + x_2^2 \quad \text{subject to} \quad x_1 x_2 = 1.$$

2. Find a solution to the problem

$$\text{Minimize} \quad -x_1^2 - x_2^2 + x_3^2$$

$$\text{subject to} \quad x_1 + x_2 = -1 \quad \text{and} \quad x_1^2 + x_2^2 = \frac{1}{2}$$

   and comment on the values of the Lagrange multipliers.
   Explain what happens if the first constraint is $x_1 + x_2 = -\frac{3}{2}$.

3. Write down the optimality conditions for Problems TD1a, TD2a and VLS1a and consider how easy they would be to solve.

16.4.   Interpretation of Lagrange multipliers

The Lagrange multipliers at the solution of a constrained optimization problem are not simply mathematical abstractions. They can be used as measures of the sensitivity of the solution with respect to changes in the constraints. Suppose that x^* solves the problem

$$\text{Minimize} \quad F(x) \quad \text{subject to} \quad c_1(x) = 0 \qquad (16.4.1)$$

and consider the *perturbed* problem

$$\text{Minimize} \quad F(x) \quad \text{subject to} \quad c_1(x) = \delta. \tag{16.4.2}$$

If the solution to (16.4.2) is $x^* + \epsilon$ then a first-order estimate of the optimum function value is

$$F(x^* + \epsilon) \approx F(x^*) + \epsilon^T \nabla F(x^*).$$

But the optimality condition for (16.4.1) states

$$\nabla F(x^*) = \lambda_1^* \nabla c_1(x^*).$$

Hence

$$F(x^* + \epsilon) \approx F(x^*) + \lambda_1 \epsilon^T \nabla c_1(x^*). \tag{16.4.3}$$

Furthermore, because $x^* + \epsilon$ solves (16.4.2), we must have

$$c_1(x^* + \epsilon) = \delta$$

and so, to the first order,

$$c_1(x^*) + \epsilon^T \nabla c_1(x^*) \approx \delta.$$

Because $c_1(x^*) = 0$ we get

$$\epsilon^T \nabla c_1(x^*) \approx \delta$$

and so (16.4.3) implies

$$F(x^* + \epsilon) - F(x^*) \approx \delta \lambda_1^*. \tag{16.4.4}$$

Hence we have shown that the Lagrange multiplier is an approximate measure of the change in the objective function that will occur if a unit amount is added to the right-hand side of the constraint. In particular, in Problem TD1a the Lagrange multiplier will indicate the extent to which the minimum surface area is changed by an increase or decrease in target volume.

The result we have just obtained generalises for problems with more than one constraint. (The proof of this is left to the reader.) Even though $\lambda_i^* \delta$ only gives an estimate of how much the objective function would change if the ith constraint were shifted by δ, these approximations are qualitatively, as well as quantitatively, useful in practice. The Lagrange multipliers with the larger magnitudes indicate which constraints have the most significant effect on the solution. Hence it would be worthwhile to try relaxing the corresponding conditions in the original problem.

Exercises

1. Extend the analysis in this section to apply to problems with more than one constraint and show that $\lambda_i^* \delta_i$ is an estimate of the change in the objective function if the ith constraint is changed to $c_i(x) = \delta_i$.

   Do a similar analysis to deal with the case when constraints i and j are shifted to become $c_i(x) = \delta_i$ and $c_j(x) = \delta_j$.

2. In the worked example in Section 16.3, use the Lagrange multiplier to predict the optimum function value when the constraint is

$$x_1 + 5x_2 - \frac{4}{3} = 0.$$

Solve the problem with this modified constraint and hence determine the accuracy of the prediction. Also do similar calculations for the modified constraint

$$x_1 + 5x_2 - \frac{2}{3} = 0.$$

16.5.   Some example problems

We now list some test problems which are used to illustrate the behaviour of constrained optimization methods. Programs implementing these problems can be downloaded along with the OPTIMA routines.

Problem TD1a is given by (16.1.1) with $V^* = 20$. We take the starting guess as $x_1 = x_2 = 2$, $x_3 = 5$ and the solution is the same as for Problem TD1. The Lagrange multiplier associated with the constraint $x_1 x_2 x_3 = 20$ is $\lambda_1^* \approx 1.17$. This enables us to estimate the minimum surface area for a different target volume. If, for example, V^* were changed to 20.5 then the minimum surface area could be expected to increase by about $0.5 \times 1.17 \approx 0.585$.

Problem TD2a is given by (16.1.2) with $S^* = 35$. We take the starting guess as $x_1 = x_2 = 2$, $x_3 = 5$ and the solution is the same as for Problem TD2. The Lagrange multiplier associated with the constraint on surface area is $\lambda_1^* \approx -0.854$.

Problem VLS1a is given by (16.1.3) with data points

$$(t_i, \ z_i) = (0, 3), \ (1, \ 8), (2, \ 12), \ (3, \ 17)$$

and the same starting guess as for the unconstrained problem VLS1. The solution is $x_1 = 3$, $x_2 \approx 4.643$. The sum-of-squares objective function in (16.1.3) has an optimum value ≈ 0.2143 which is greater than the minimum function value when the equality constraint is not present.

Problem VLS2a is given by (16.1.4) with the data points

$$(t_i, \; z_i) = (0, 1), \; (1, \; 0.5), (2, \; 0.4), \; (3, \; 0.3), \; (4, \; 0.2)$$

and the same starting guess as for Problem VLS2. Figure 16.1 shows the contours of the function with a dotted line indicating the constraint. The solution of the unconstrained problem VLS2 (marked with an asterisk) is infeasible and the constrained solution is marked by a circle at $x_1 \approx 0.926$, $x_2 \approx -0.383$.

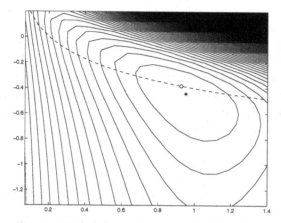

Figure 16.1.   Contour plot for Problem VLS2a.

Problem TLS1a is given by (16.1.6), (16.1.7) with data points

$$(t_i, \; z_i) = (0, 3), \; (1, \; 8), (2, \; 12), \; (3, \; 17).$$

The starting guess has $x_1 = \cdots = x_6 = 0$. The solution values for the parameters in the model function are $x_1 \approx 3.087$, $x_2 \approx 4.608$. The footpoints are

$$-0.0181, \; 1.063, \; 1.937, \; 3.018$$

and the sum-of-squares error function (16.1.6) is approximately 0.009.

Problem TLS2a is given by (16.1.8), (16.1.9) with the data points

$$(t_i, \; z_i) = (0, 1), \; (1, \; 0.5), (2, \; 0.4), \; (3, \; 0.3), \; (4, \; 0.2)$$

and the starting guess $x_1 = \cdots = x_7 = 0$. The solution gives the model function

$$z = x_1 e^{x_2 t} \quad \text{with} \quad x_1 \approx 0.9486, \quad x_2 \approx -4.406$$

and the footpoints, τ_i, are approximately

$$-0.0184, \quad 1.027, \quad 1.999, \quad 2.995, \quad 3.997.$$

The sum-of-squares error function is approximately 0.0173.

Problems OC1a(n) and OC2a(n) are given, respectively, by (16.1.10) and (16.1.11). We use the values

$$\tau = \frac{3}{n}, \quad u_0 = u_f = 0, \quad s_0 = 0 \quad \text{and} \quad s_f = 1.5.$$

The starting guess is the same as for the corresponding unconstrained problems OC1(n) and OC2(n). Solutions differ from those of OC1 chiefly in the fact that the terminal conditions $u_f = 0$, $s_f = 1.5$ are satisfied exactly. At the solution to OC1, the errors in the final speed and position are simply included as part of the overall objective function and the unconstrained minimization of (3.3.2) does not force them to zero. Thus, in the case $n = 4$ the solution of OC1 is given by

$$x_1 \approx 0.746, \quad x_2 \approx 0.378, \quad x_3 \approx -0.362, \quad x_4 \approx -0.735$$

giving $u_f \approx 0.02$, $s_f \approx 1.49$. The solution of OC1a on the other hand is

$$x_1 \approx 0.762, \quad x_2 \approx 0.381, \quad x_3 \approx -0.381, \quad x_4 \approx -0.762$$

$$\text{giving} \quad u_f = 0, \quad s_f = 1.5.$$

(The reader can compare the solutions of OC2a with those of OC2.)

Problem OC3(n) is defined by (16.1.12)–(16.1.16) using the same definitions of τ, u_0, u_f and s_f as in OC1a(n) and OC2a(n). The value of c_D is taken as 0.1. For the case when $n = 4$, Table 16.1 shows how the quadratic drag term affects the solution of OC3 as c_D increases from zero. The reader is invited to explain why the symmetry in the optimal accelerations when $c_D = 0$ (i.e., $x_4 = -x_1$ and $x_3 = -x_2$) is not maintained when c_D is nonzero.

c_D	x_1	x_2	x_3	x_4	F^*
0.0	0.762	0.381	−0.381	−0.762	2.032
0.033	0.773	0.401	−0.367	−0.764	2.069
0.067	0.785	0.421	−0.353	−0.767	2.108
0.1	0.797	0.441	−0.339	−0.769	2.147

Table 16.1.   Solutions of Problem OC3(4) for varying c_D.

Problem FBc is given by (16.1.17) and (16.1.18). The function $\psi(x)$ is defined by (15.3.5) with $x_{min} = 1$. We use $n = 3$ and $m = 1$ and the data values

$$c_1 = 1.5, \quad c_2 = 1, \quad c_3 = 0.8;$$

$$\alpha_{11} = 0.12, \quad \alpha_{12} = 0.08, \quad \alpha_{13} = 0.06; \quad \beta_1 = 3.75.$$

The starting guess is $x_1 = x_2 = x_3 = 17$. There are (at least) three local solutions:

$$x_1 \approx 0, \quad x_2 \approx 37.5, \quad x_3 \approx 12.5 \quad \text{giving} \quad C \approx 47.5;$$

$$x_1 \approx 1, \quad x_2 \approx 34.5, \quad x_3 \approx 14.5 \quad \text{giving} \quad C \approx 47.6;$$

$$x_1 \approx 12.5, \quad x_2 \approx 0, \quad x_3 \approx 37.5 \quad \text{giving} \quad C \approx 48.75.$$

The methods described in the chapters which follow are all local optimization techniques and can be said to have behaved acceptably if they converge to any one of these solutions. (The feed-blending problem is one example of a larger class of *resource allocation* problems in which demands have to be met by the most economical use of several sources of supply. Another instance would be the optimal use of several generating sets to meet a known demand in an electrical supply network.)

We conclude this section by noting that we do not use R1(2)c as a practical example because it is a poorly-posed problem for gradient-based methods. The constraint functions in (16.1.20) are identically zero for any route which does not pass through the no-go region and therefore all the first derivatives of ν are also identically zero for any feasible point. If any of the $\nabla\nu = 0$ then the constraint normals cannot be linearly independent. This will lead to difficulties in the calculation or estimation of Lagrange multipliers. A further drawback for gradient-based methods of constrained optimization is that the first derivatives of ν are discontinuous at the boundaries of the no-go region.

Loss adjusters (Part 2) [5]

Partnership or kinship means
they share a common blueprint. One's left-handed
so they sidle counter crab-wise
scavenging round tragedies.
Beyond too late, there's always time
for lodging ever-overdue objections
to tenders that should not have won the contract.

The signals defaulting to safe not to danger;
no lightning rod earthing the main mooring mast;
not enough lifeboats for all the ship's complement;
the iron bridge girders imperfectly cast.

Chapter 17

Linear Equality Constraints

17.1.   Quadratic programming

We consider first the special case when the function (16.2.1) is quadratic and the constraints (16.2.2) are linear. The problem is then an equality constrained *quadratic programming problem* (EQP). It can be written as

$$\text{Minimize } \frac{1}{2}(x^T G x) + h^T x + c \quad \text{subject to } Ax + b = 0 \qquad (17.1.1)$$

where the $n \times n$ matrix G and the $l \times n$ matrix A are both constant. The first-order optimality conditions for (17.1.1) are

$$Ax^* + b = 0 \qquad (17.1.2)$$

$$Gx^* + h - A^T \lambda^* = 0. \qquad (17.1.3)$$

After rearrangement, these become a system of $n + l$ linear equations

$$\begin{pmatrix} G & -A^T \\ -A & 0 \end{pmatrix} \begin{pmatrix} x^* \\ \lambda^* \end{pmatrix} = \begin{pmatrix} -h \\ b \end{pmatrix}. \qquad (17.1.4)$$

One way of solving (17.1.1) is simply to form the linear system (17.1.4) and find x^*, λ^* using (say) Gaussian elimination. (Although (17.1.4) is symmetric, the zeros on the diagonal imply that it is not positive-definite and so the Cholesky method is not suitable.)

If G is positive-definite then the feasible stationary point obtained from (17.1.4) will be a minimum. Otherwise we must check the second-order condition

$$z^T G z > 0 \quad \text{for all } z \quad \text{s.t. } Az = 0,$$

to confirm that x^* is not a constrained maximum or saddle point.

M. Bartholomew-Biggs, *Nonlinear Optimization with Engineering Applications*,
DOI: 10.1007/978-0-387-78723-7_17, © Springer Science+Business Media, LLC 2008

Forming and solving the whole system of equations (17.1.4) may be computationally efficient when the matrices G and A are sparse. However, we can also find x^* and λ^* separately. For instance, if we multiply (17.1.3) by AG^{-1} and then use (17.1.2) to eliminate x^*, we can get λ^* from

$$(AG^{-1}A^T)\lambda^* = AG^{-1}h - b. \qquad (17.1.5)$$

It then follows that

$$Gx^* = A^T\lambda^* - h. \qquad (17.1.6)$$

Solving (17.1.5) and (17.1.6) separately costs $O(l^3) + O(n^3)$ multiplications. This is less than the $O((n+l)^3)$ multiplications needed to solve (17.1.4). However, we must also allow for the cost of forming $AG^{-1}A^T$. This matrix product takes $l^2n + ln^2$ multiplications. Moreover, the inversion of G takes $O(n^3)$ multiplications (but then we can use G^{-1} to avoid the cost of solving (17.1.6) from scratch). This fuller analysis suggests that there may be little computational advantage in using (17.1.5), (17.1.6) except when G^{-1} is already known. In this case the solution cost is $O(l^3 + l^2n + ln^2)$ multiplications which is much less than $O((n+l)^3)$, especially when $l \ll n$.

17.2.   Sample EQP solutions

Let us consider the constrained data-fitting problem similar to VLS1a

$$\text{Minimize} \sum_{i=2}^{4}(x_1 + x_2t_i - z_i)^2 \quad \text{s.t.} \quad x_1 + x_2t_1 - z_1 = 0 \qquad (17.2.1)$$

where the data points (t_i, z_i) are (0,5), (1,8), (2,12), (3,16).
    On substituting the data values, the objective function becomes

$$F(x) = (x_1 + x_2 - 8)^2 + (x_1 + 2x_2 - 12)^2 + (x_1 + 3x_2 - 16)^2$$

which simplifies to

$$F(x) = 3x_1^2 + 14x_2^2 + 12x_1x_2 - 72x_1 - 160x_2 + 464.$$

Hence the objective function in (17.2.1) can be written as

$$\frac{1}{2}x^TGx + h^Tx + c$$

where

$$G = \begin{pmatrix} 6 & 12 \\ 12 & 28 \end{pmatrix}, \quad h = \begin{pmatrix} -72 \\ -160 \end{pmatrix} \quad \text{and} \quad c = \sum_{i=2}^{4} z_i^2 = 464.$$

The constraint is $x_1 + x_2 t_1 = z_1$ and so the Jacobian is

$$A = (1, \ t_1).$$

On substituting the data values, the constraint is $x_1 - 5 = 0$ and the Jacobian is $A = (1, 0)$. Therefore the optimality conditions give (17.1.4) as

$$\begin{pmatrix} 6 & 12 & 1 \\ 12 & 28 & 0 \\ 1 & 0 & 0 \end{pmatrix} \begin{pmatrix} x_1^* \\ x_2^* \\ \lambda^* \end{pmatrix} = \begin{pmatrix} 72 \\ 160 \\ 5 \end{pmatrix}.$$

This system is easy to solve. We get $x_1^* = 5$ immediately from the third equation and then, from the second equation, we obtain

$$x_2^* = \frac{1}{28}(160 - 12x_1^*) = \frac{100}{28} = \frac{25}{7}.$$

Finally we can find λ^* from the first equation as

$$\lambda^* = 72 - 6x_1^* - 12x_2^* = -\frac{6}{7}. \tag{17.2.2}$$

As a second example we consider (16.3.1) which is a quadratic programming problem that can be written as

$$\text{Minimize} \quad \frac{1}{2}x^T \begin{pmatrix} 2 & 3 \\ 3 & 0 \end{pmatrix} x \quad \text{s.t.} \ (1 \ 5)x - 1 = 0.$$

Hence, in the notation of 17.1.1),

$$G = \begin{pmatrix} 2 & 3 \\ 3 & 0 \end{pmatrix}, \quad h = \begin{pmatrix} 0 \\ 0 \end{pmatrix}, \quad c = 0, \quad A = (1 \ 5), \quad b = -1.$$

We can use the solution method (17.1.5), (17.1.6). Because

$$G^{-1} = \begin{pmatrix} 0 & 1/3 \\ 1/3 & -2/9 \end{pmatrix}$$

we obtain

$$AG^{-1}A^T = (1 \ 5)^T \begin{pmatrix} 0 & 1/3 \\ 1/3 & -2/9 \end{pmatrix} \begin{pmatrix} 1 \\ 5 \end{pmatrix} = -\frac{20}{9}$$

and $\quad AG^{-1}h - b = 1.$

Hence, from (17.1.5), the Lagrange multiplier is

$$\lambda^* = (AG^{-1}A^T)^{-1}(AG^{-1}h - b) = -\frac{9}{20}.$$

From (17.1.6), the optimal x is given by

$$x^* = G^{-1}(A^T\lambda^* - h) = \begin{pmatrix} 0 & 1/3 \\ 1/3 & -2/9 \end{pmatrix} \begin{pmatrix} -9/20 \\ -9/4 \end{pmatrix} = \begin{pmatrix} -3/4 \\ 7/20 \end{pmatrix}.$$

This agrees with the solution obtained in Section 16.3.

Exercises

1. Evaluate the function in (17.2.1) at the solution $x_1 = 5$, $x_2 = 25/7$. Use the Lagrange multiplier in (17.2.2) to estimate the optimal objective function value if the point (t_1, z_1) is changed to $(0, 6)$. Compare your result with the actual solution obtained by forming and solving (17.1.4).
2. Use the Lagrange multiplier in (17.2.2) to estimate the function value at the solution of (17.2.1) when the first data point is $(0, 4)$. Show that the actual solution to this problem has a zero function value and explain why the Lagrange multiplier is also zero.
3. Use (17.1.5) and (17.1.6) to solve the problem

   Minimize $\quad x_1^2 + 3x_2^2 - 2x_1 - 4x_2 + 6 \quad$ s.t. $\quad 7x_1 - 3x_2 + 10 = 0.$

4. Form and solve an extended version of problem (17.2.1) which involves two more data points $(4, 21)$ and $(5, 25)$ and in which the model function is $z = x_1 + x_2 t + x_3 t^2$.

17.3.   Reduced-gradient methods

Reduced-gradient methods use linear equality constraints to eliminate some of the variables. The linear constraints in (17.1.1) are

$$Ax + b = 0$$

and we suppose that A is partitioned as $(\tilde{A} : \bar{A})$ where $\tilde{A}$ has $n - l$ columns and $\bar{A}$ has l columns. If we similarly partition the vector of variables as $(\tilde{x} : \bar{x})$ then the constraints can be written

$$\tilde{A}\tilde{x} + \bar{A}\bar{x} = b.$$

Hence, if $\bar{A}$ is nonsingular,

$$\bar{x} = \bar{A}^{-1}(b - \tilde{A}\tilde{x}) \qquad\qquad (17.3.1)$$

and we have expressed the last l variables in terms of the first $n-l$. This means we can write

$$x = v + M\tilde{x}$$

where v is a column vector of length n and M is an $n \times (n-l)$ matrix. These are given, in partitioned form, by

$$v = \left(\begin{array}{c} 0_{(n-l) \times 1} \\ -\,-\,- \\ \bar{A}^{-1}b \end{array} \right); \quad M = \left(\begin{array}{c} I_{(n-l) \times (n-l)} \\ -\,-\,- \\ -\bar{A}^{-1}\tilde{A}) \end{array} \right).$$

Substituting in the objective function in (17.1.1) we get

$$\tilde{F}(\tilde{x}) = \frac{1}{2}(\tilde{x}^T M^T + v^T)G(v + M\tilde{x}) + h^T(v + M\tilde{x}) + c$$

which simplifies to

$$\tilde{F}(\tilde{x}) = \frac{1}{2}\tilde{x}^T(M^T G M)\tilde{x} + (v^T GM + h^T M)\tilde{x} + v^T Gv + h^T v + c.$$

Thus solving (17.1.1) has been reduced to finding the unconstrained minimum of the $(n-l)$-variable quadratic function $\tilde{F}$ in terms of the variables $\tilde{x}$.

As a simple illustration of the reduced-gradient approach we consider problem (16.3.1) and use the constraint to eliminate x_1. Because $x_1 = 1 - 5x_2$ we can rewrite the objective function in terms of x_2 only and obtain

$$\tilde{F} = x_1^2 + 3x_1 x_2 = 1 - 10x_2 + 25x_2^2 + 3x_2 - 15x_2^2 = 10x_2^2 - 7x_2 + 1.$$

We can find the minimum by setting $d\tilde{F}/dx_2 = 0$ which gives $x_2 = 7/20$. Hence $x_1 = 1-35/20 = -3/4$ and (of course) we obtain the same solution as that given by previous approaches.

Exercises

1. Solve problem (16.3.1) by the reduced-gradient method based on using the constraint to eliminate x_2.

2. Solve the problem

$$\text{Minimize} \quad x^T Q x$$

$$\text{s.t.} \quad 0.2x_1 + 0.3x_2 + 0.15x_3 = 0.2 \quad \text{and} \quad x_1 + x_2 + x_3 = 1.$$

where

$$Q = \left(\begin{array}{ccc} 0.06 & -0.02 & 0 \\ -0.02 & 0.05 & 0 \\ 0 & 0 & 0 \end{array} \right)$$

and use the reduced-gradient approach to transform it to an unconstrained problem involving x_1 only.

3. Use both (17.1.5), (17.1.6) and the reduced gradient approach to find a solution to

$$\text{Minimize} \quad x_1 x_2 \quad \text{s.t.} \quad x_1 - x_2 = 1$$

and confirm that the point obtained is a constrained minimum. What happens if you apply the reduced-gradient method to

$$\text{Minimize} \quad x_1 x_2 \quad \text{s.t.} \quad x_1 + x_2 = 1?$$

Range- and null-spaces

The reduced-gradient approach can be implemented in a more general way that does not require us to partition the constraint matrix to find a nonsingular $\bar{A}$. Instead we consider a solution in terms of its components in two subspaces which are *normal* and *tangential* to the constraints.

The optimality conditions (16.2.7) and (16.2.8) in Chapter 16 involve feasible directions which lie in the tangent space of the constraints. If an n-variable optimization problem involves l linear constraints, $Ax + b = 0$, then a feasible direction can be any member of the $(n - l)$-dimensional subspace of vectors z which satisfy $Az = 0$.

Definition The $(n - l)$-dimensional subspace of vectors z giving $Az = 0$ is called the *null-space* of A.

We now let Z be an $n \times (n - l)$ matrix whose columns span the null-space of A. This means that Zw is a feasible direction for any $(n - l)$-vector. The choice of Z is not unique: but one way of obtaining it is by *orthogonal factorization* of the constraint Jacobian A. This factorization (see [38] for more details) yields an *orthonormal* $n \times n$ matrix Q and an $l \times l$ lower triangular matrix L such that

$$AQ = R = (L : 0) \quad \text{and} \quad Q^T Q = I. \tag{17.3.2}$$

If we now let Y be the matrix composed of the first l columns of Q and Z the matrix consisting of the remaining $(n - l)$ columns then it can be shown that

$$AZ = 0, \quad AY = L \quad \text{and} \quad Y^T Z = 0. \tag{17.3.3}$$

Definition The l-dimensional subspace spanned by the columns of Y is called the *range-space* of the constraint Jacobian A.

Definition $Z^T \nabla F(x)$ and $Z^T \nabla^2 F(x) Z$ are called, respectively, the *reduced gradient* and *reduced Hessian* of $F(x)$.

The reduced-gradient approach can be based on considering separately the components of the solution x^* which lie in the range- and null-spaces of the constraint Jacobian. In the next subsection we show how this is done for equality constrained QP problems. Subsequently we describe the method for more general linearly constrained problems.

Exercise
Use (17.3.2) to verify the relationships in (17.3.3).

Reduced gradients and EQP

If x^* solves (17.1.1) we can find its components in Y- and Z-space. Suppose that $\bar{y}$ is an l-vector and $\bar{z}$ is an $(n-l)$-vector such that

$$x^* = Y\bar{y} + Z\bar{z}. \tag{17.3.4}$$

The components $Y\bar{y}$ and $Z\bar{z}$ are sometimes called the *vertical step* and the *horizontal step*. Because $AZ = 0$, optimality condition (17.1.2) implies

$$AY\bar{y} + b = 0. \tag{17.3.5}$$

From (17.3.3) we get $AY = L$ and this means that $\bar{y}$ can be found by solving a lower triangular system of equations.

On premultiplying optimality condition (17.1.3) by Z^T we get

$$Z^T G Z \bar{z} = -Z^T h - Z^T G Y \bar{y} \tag{17.3.6}$$

which is a symmetric system of equations for $\bar{z}$. If the EQP has a minimum then the reduced Hessian $Z^T G Z$ is positive-definite and so the Cholesky method can be used to solve (17.3.6).

If we premultiply (17.1.3) by Y^T we get an upper triangular system

$$Y^T A^T \lambda^* = Y^T h + Y^T G x^* \tag{17.3.7}$$

which can be solved to give the Lagrange multipliers.

Solving (17.3.5), (17.3.6) and (17.3.7) requires $O(l^2) + O((n-l)^3)$ multiplications which can be appreciably less than $O((n+l)^3)$ multiplications needed to solve the system (17.1.4) – especially when $l \approx n$. This comparison, however, neglects the cost of finding the Y and Z basis matrices.

We can illustrate the above approach by considering problem (16.3.1) again. Recall from Section 17.2 that this problem is a standard EQP of the form (17.1.1) with

$$G = \begin{pmatrix} 2 & 3 \\ 3 & 0 \end{pmatrix}, \quad h = \begin{pmatrix} 0 \\ 0 \end{pmatrix}, \quad c = 0, \quad A = (1 \ 5), \quad b = -1.$$

We can show that the matrices

$$Y = \begin{pmatrix} 1/\sqrt{26} \\ 5/\sqrt{26} \end{pmatrix} \quad Z = \begin{pmatrix} 5/\sqrt{26} \\ -1/\sqrt{26} \end{pmatrix}$$

satisfy the conditions

$$AZ = 0, \quad AY = \sqrt{26}, \quad Y^T Z = 0, \quad Y^T Y = Z^T Z = 1.$$

Hence Y and Z span the range- and null-space of the constraint Jacobian A.

From (17.3.5)

$$AY\bar{y} + b = 0 \quad \text{implies} \quad \sqrt{26}\bar{y} - 1 = 0$$

and so $\bar{y} = 1/\sqrt{26}$. Hence

$$Y\bar{y} = \begin{pmatrix} 1/26 \\ 5/26 \end{pmatrix}.$$

We also have

$$Z^T G = (7/\sqrt{26}, \quad 15/\sqrt{26})$$

and so

$$Z^T GZ = 20/26 \quad \text{and} \quad Z^T GY\bar{y} = 82/(26\sqrt{26}).$$

Therefore, from (17.3.6),

$$Z^T GZ\bar{z} = -Z^T h - Z^T GY\bar{y} \quad \text{implies} \quad \frac{20}{26}\bar{z} = -\frac{82}{26\sqrt{26}}.$$

Hence

$$\bar{z} = -\frac{41}{10\sqrt{26}} \quad \text{and} \quad Z\bar{z} = \begin{pmatrix} -41/52 \\ 41/260 \end{pmatrix}.$$

We now have

$$x^* = Y\bar{y} + Z\bar{z} = \begin{pmatrix} -39/52 \\ 91/260 \end{pmatrix} = \begin{pmatrix} -3/4 \\ 7/20 \end{pmatrix}.$$

Once again we have obtained the same optimal values of the variables as those found in Section 16.3. In order to obtain the Lagrange multiplier from (17.3.7) we calculate

$$Y^T G = (1/\sqrt{26} \quad 5/\sqrt{26}) \begin{pmatrix} 2 & 3 \\ 3 & 0 \end{pmatrix} = (17/\sqrt{26}, \quad 3/\sqrt{26})$$

and so

$$Y^T G x^* = -117/(10\sqrt{26}).$$

Then (17.3.7) becomes

$$\sqrt{26}\lambda^* = -117/(10\sqrt{26}) \quad \text{giving} \quad \lambda^* = -117/260 = -9/20$$

which also agrees with previous results.

Exercise
Consider the two-variable problem

$$\text{Minimize} \quad x^T M x \quad \text{s.t.} \quad x_1 + x_2 = 1. \tag{17.3.8}$$

The constraint Jacobian A is then the row-vector $(1,1)$. Show that $AQ = (l_{11}, 0)$ when Q is a 2×2 matrix defined by

$$Q = I - 2\frac{ww^T}{w^T w} \quad \text{with} \quad w = \begin{pmatrix} 1 + \sqrt{2} \\ 1 \end{pmatrix}$$

and hence find l_{11}. Use (17.3.5)–(17.3.7) to solve (17.3.8).

General linearly constrained problems

The reduced-gradient approach can be applied to (16.2.1), (16.2.2) when the constraints are linear but the function is nonquadratic and the problem is of the form

$$\text{Minimize} \quad F(x) \quad \text{subject to} \quad Ax + b = 0. \tag{17.3.9}$$

As in unconstrained optimization, a common strategy is to use an iterative scheme based on local quadratic approximations to F. That is, in the neighbourhood of a solution estimate, x, we suppose

$$F(x + p) \approx Q(p) = F + p^T \nabla F + \frac{1}{2}(p^T B p) \tag{17.3.10}$$

where either $B = \nabla^2 F(x)$ or $B \approx \nabla^2 F(x)$. The following algorithm solves problem (17.3.9) by using the quadratic approximation (17.3.10) to generate a search direction on each iteration. For reasons explained below, it is based on using an updated approximation to the Hessian matrix rather than the exact $\nabla^2 F(x)$.

Reduced-gradient algorithm for linear equality constraints

Choose an initial feasible point x_0 and set $\lambda_0 = 0$
Choose B_0 as a positive definite estimate of $\nabla^2 F(x_0)$.

Obtain Y and Z as basis matrices for the range and null spaces of A
Repeat for $k = 0, 1, 2, \ldots$
Set $g_k = \nabla F(x_k)$
  Determine $\bar{z}$ from $Z^T B_k Z \bar{z} = -Z^T g_k$ and set $p_k = Z\bar{z}$
  Obtain λ_{k+1} by solving $Y^T A^T \lambda = Y^T g_k + Y^T B_k p_k$
  Perform a line search to get $x_{k+1} = x_k + sp_k$ where $F(x_{k+1}) < F(x_k)$
  Obtain B_{k+1} from B_k by a quasi-Newton update (see below).
until $\|Z^T g_k\|$ is less than a specified tolerance.

This algorithm proceeds in a similar way to the reduced-gradient method for an EQP. Each iteration makes a "horizontal" move in the subspace satisfying the constraints. No "vertical" move is needed because the algorithm described above is a *feasible point technique*. This means that it must be provided with a feasible guessed solution and then all subsequent iterates will also be feasible. (In practice, the algorithm could be preceded by some initial iterations to find a feasible point.)

The algorithm includes a line search because the nonquadraticity of F means that $x + p$ is not guaranteed to be a "better" point than x. The algorithm can use either a perfect line search to minimize $F(x + sp)$ or a weak line search to ensure that $F(x + sp) - F(x)$ is negative and satisfies the Wolfe conditions.

The advantage of using B_k as an updated estimate of $\nabla^2 F$ is that it enables us to keep B_k positive-definite. This will ensure that the matrix $Z^T B_k Z$ is positive-definite and hence that p_k is a descent direction on every iteration.

The BFGS update (Chapter 10) generates matrices B_{k+1} as successive estimates of $\nabla^2 F$ which satisfy the quasi-Newton condition (10.5.2). These estimates are all positive-definite provided (10.2.2) is satisfied. There are also modified updating formulae which yield positive-definite B even when $\delta_k^T \gamma_k \leq 0$. One such (called Powell's modification [52]) involves replacing the actual change in gradient, γ_k, in (10.5.3) by

$$\eta_k = (1 - \theta)\gamma_k + \theta B_k \delta_k \qquad (17.3.11)$$

with θ being chosen so that $\delta_k^T \eta_k > 0$.

In the unconstrained case we can justify the use of updates which force B_k to be positive-definite because we know that $\nabla^2 F(x^*)$ must be positive-definite. In constrained problems, however, the true Hessian $\nabla^2 F(x^*)$ is often indefinite. This suggests that a positive-definite updating scheme may be inconsistent with making B_k a good approximation. In fact there is no conflict of interest because the second-order optimality condition (16.2.8) only relates to the null space of the binding constraint normals. In this subspace the optimality conditions require the Hessian to be positive-definite.

It is of course possible to implement reduced-gradient methods which use exact second derivatives, but then the search direction calculation must be adapted to ensure that a descent direction is still obtained even when $\nabla^2 F(x)$ is indefinite.

Proofs of convergence for reduced-gradient algorithms can be based on ideas already discussed in relation to unconstrained minimization algorithms. Under fairly mild assumptions about the functions and constraints and about the properties of the Hessian (or its updated estimate) it can be shown that the search directions and step lengths in the feasible subspace will satisfy the Wolfe conditions. It then follows that the iterations will converge to a point where the reduced gradient is zero. The ultimate rate of convergence can be superlinear (or quadratic if exact second derivatives are used instead of updated approximations to the Hessian).

17.4.   Results with a reduced-gradient method

A powerful reduced-gradient method, known as GRG (see Lasdon *et al.* [43]) is implemented in the SOLVER tool in Microsoft Excel [29, 48]. When applied to a linearly constrained problem, the GRG method works in a way that is broadly similar to the algorithm described in section 17.3 above. The SOLVER implementation uses approximate derivative information and so it will not usually deal with an EQP in a single iteration because the Hessian matrix is not available for use in equations (17.1.4). SOLVER can be used on problems where the function is not quadratic and – unlike the algorithm in section 17.3 above – it can be started from an infeasible point. As we show in later chapters, SOLVER can also deal with nonlinear constraints.

We now apply SOLVER to the linearly constrained example problems VLS1a, OC1a, OC2a and FBc. The first two are quadratic programming problems but the second two have nonquadratic objective functions. Table 17.1 shows the number of iterations used by SOLVER. The figure in brackets is the number of iterations needed to obtain a feasible point. (The software does not report the number of function evaluations used.) We note that, for these problems, the number of iterations is roughly the same as the number of variables except in the case of the highly nonconvex problem FBc.

VLS1a	OC1a(4)	OC1a(6)	OC2a(4)	OC2a(6)	FBc
3(1)	3(1)	5(1)	5(1)	7(1)	9(2)

Table 17.1.   Iteration counts for SOLVER with linear constraints.

Exercises

1. Write an implementation (e.g., a spreadsheet calculation or a MATLAB script) of a solution method for an EQP of the form (17.1.1). Given the matrices and vectors G, h, A, b it should form and solve the system (17.1.4).

   Apply your implementation to Problems VLS1a and OC2a(4) and show that the same results are obtained if (17.1.5) and (17.1.6) are used instead of (17.1.4).

2. Apply a reduced-gradient method to problems similar to OC1a and OC2a using the data $t_f = 5$, $\tau = t_f/n$, $u_0 = s_0 = 0$, $s_f = 3$ with no restriction on u_f.

3. By using different starting guesses, see how many local solutions you can find for problem FBc.

4. Form an extended version of Problem VLS1a which involves two more data points $(4, 21)$ and $(5, 25)$ and uses the quadratic model function $z = x_1 + x_2 t + x_3 t^2$. Use SOLVER to obtain a solution.

17.5.   Projected-gradient methods

If x_k is a feasible point for the linearly constrained problem

$$\text{Minimize}\quad F(x)\quad \text{subject to}\quad Ax + b = 0.$$

then we can obtain a search direction by projecting a descent direction for F into the feasible subspace. For instance,

$$p_k = -(I - A^T(AA^T)^{-1}A)g_k \qquad (17.5.1)$$

is a projection of the negative gradient $-g_k$. The matrix $P = I - A^T(AA^T)^{-1}A$ is called a *projection matrix* and it is easy to show that $AP = 0$. Hence the search direction (17.5.1) satisfies $Ap_k = 0$ and a new point

$$x_{k+1} = x_k + sp_k$$

will be feasible if x_k is feasible. Given an initial feasible point x_0, we can use line searches along directions given by (17.5.1) for $k = 0, 1, 2, \ldots$ in order to minimize F in the feasible subspace.

The projected-gradient (17.5.1) is, in general, no more efficient than the steepest descent direction for unconstrained minimization. However, we can also obtain projections of more effective descent directions. If $B \approx \nabla^2 F(x_k)$ is a positive definite matrix then

$$p_k = -B^{-1}(I - A^T(AB^{-1}A^T)^{-1}A)B^{-1}g_k \qquad (17.5.2)$$

is a projected quasi-Newton direction.

Clearly, projected-gradient approaches work in much the same way as reduced-gradient methods, by restricting the search to the feasible subspace. Reduced gradients have a practical advantage for large-scale problems, however, because some of their algebra involves $(n-l) \times (n-l)$ matrices, whereas the projection methods use $n \times n$ matrices throughout.

Exercises

1. If x_k is a feasible point for EQP (17.1.1), obtain an expression for the step p such that $x_k + p$ is the solution. By writing the expression for p in the form (17.1.6), show that p can also be viewed as a projected Newton direction.

2. Write an algorithm which uses projected quasi-Newton directions to minimize $F(x)$ subject to linear constraints $Ax + b = 0$.

Loss adjusters (Part 3) [5]

Why make attempts to make amends
for other parties' negligence or crimes?
After blaming's had its day
in court, no praise awaits portfolios
of hindsights. Sorting should-have-beens,
to salvage just one could-be: this, they must
believe, does more than set a record straight.

No missing bulkheads to weaken the vessel;
fill no more airships with porous gas-bags;
add reinforcement at corners of windows;
let cracks be acknowledged, not hidden by flags.

Chapter 18

Penalty Function Methods

18.1. Introduction

We now turn to methods for dealing with nonlinear constraints in problem (16.2.1), (16.2.2). These are usually considered to present more difficulties than nonquadraticity in the objective function. This is largely because it is hard to ensure all iterates remain feasible. Hence the main focus of this chapter is on methods which do not generate feasible points on every iteration but merely force the solution estimates x_k to approach feasibility as they converge. We begin, however, by considering the extension of the reduced-gradient approach to deal with nonlinear equality constraints.

Reduced-gradients and nonlinear constraints

The reduced-gradient method, described in Section 17.3, can be extended to deal with nonlinear constraints. The chief difficulty to be overcome is that of maintaining feasibility because a step along a horizontal search direction p does not now ensure that $c_i(x + sp) = 0$ for each constraint. Thus we need a *restoration* strategy in which a basic horizontal move is followed by a vertical step back onto the constraints. A first estimate of this restoration step can be obtained by defining

$$\hat{c}_i = c_i(x + sp) \quad \text{for} \quad i = 1, \dots, l$$

then finding y to solve $AYy = -\hat{c}$ and finally setting $\hat{p} = Yy$. If the constraints (16.2.2) are near-linear then the point

$$x^+ = x + sp + \hat{p}$$

M. Bartholomew-Biggs, *Nonlinear Optimization with Engineering Applications*,
DOI: 10.1007/978-0-387-78723-7_18, © Springer Science+Business Media, LLC 2008

may be near-feasible and suitable for the start of a new iteration. However, when the c_i are highly nonlinear the calculation of a suitable restoration step may itself be an iterative process.

Another aspect of the reduced-gradient algorithm that must be modified when dealing with nonlinear constraints concerns the Hessian matrix in the local quadratic model (17.3.10). The second-order optimality condition for nonlinearly constrained problems is (16.2.9) which involves the Hessian of the Lagrangian rather than the objective function. This means that, in the discussion of the reduced gradient algorithm in Section 17.3, the matrix B should be regarded as an approximation to $\nabla^2 L^*$ where

$$\nabla^2 L^* = \nabla^2 F - \sum_{i=1}^{l} \lambda_i^* \, \nabla^2 c_i.$$

If the matrix B_k in the reduced-gradient algorithm is to be calculated via a quasi-Newton approach then a suitable update can be obtained by redefining γ_k in the quasi-Newton condition $B_{k+1}\delta_k = \gamma_k$ as

$$\gamma_k = \nabla L(x_{k+1}) - \nabla L(x_k)$$

where L is a local approximation to L^* based on Lagrange multiplier estimates, λ_{k_i}, determined at x_k. Hence

$$\nabla L(x) = \nabla F(x) - \sum_{i=1}^{l} \lambda_{k_i} \, \nabla c_i(x).$$

If $\delta_k^T \gamma_k$ is not positive then we can use (17.3.11) to define η_k as a replacement for γ_k so that the BFGS update will make B_{k+1} positive-definite.

Numerical results with SOLVER

Table 18.1 summarises the performance of the SOLVER implementation of the reduced-gradient method when applied to some nonlinearly constrained problems. As in Table 17.1, the entries are numbers of iterations needed for convergence with a bracketed figure showing how many iterations are needed to obtain feasibility.

TD1a	TD2a	VLS2a	TLS1a	TLS2a	OC3(6)
5(0)	7(1)	20(4)	62(3)	14(4)	12(7)

Table 18.1.   Iteration counts for SOLVER with nonlinear constraints.

Results for TLS1a illustrate the practical difficulties that sometimes occur when the reduced-gradient method is used with nonlinear constraints. SOLVER locates a feasible point in just 3 iterations but, unfortunately, this point is not very close to the optimum and so a further 59 iterations are needed during which the search has to work its way round the curved constraint until it reaches the constrained minimum. This behaviour also occurs in a less extreme form on Problems VLS2a and TLS2a. Because the reduced-gradient approach may make slow progress around curved constraints it is worth considering methods which do not depend on maintaining feasibility on every iteration.

18.2. Penalty functions

We can avoid the difficulties of maintaining feasibility with respect to nonlinear constraints by converting (16.2.1), (16.2.2) into a sequence of unconstrained problems.

Definition A *penalty function* associated with (16.2.1), (16.2.2) is

$$P(x, \ r) = F(x) + \frac{1}{r} \sum_{i=1}^{l} c_i(x)^2 \quad \text{where} \quad r > 0. \tag{18.2.1}$$

The quantity r is called the *penalty parameter*. When x is a feasible point, $P(x, r) = F(x)$. When x is infeasible then P exceeds F by an amount proportional to the square of the constraint violations. An important property of the penalty function (18.2.1) is as follows.

Proposition Suppose that, in the problem (16.2.1), (16.2.2), $F(x)$ is bounded below for all x and that there is a unique solution x^* where the constraint normals $\nabla c_1(x^*), \ldots, \nabla c_l(x^*)$ are linearly independent. Suppose also that ρ is positive and that, for all $r_k < \rho$, the Hessian matrix $\nabla^2 P(x, r_k)$ is positive-definite for all x. Then if x_k solves the unconstrained problem

$$\text{Minimize } P(x, r_k) \tag{18.2.2}$$

it follows that

$$x_k \to x^* \quad \text{as} \quad r_k \to 0 \tag{18.2.3}$$

and also

$$-\frac{2c_i(x_k)}{r_k} \to \lambda_i^* \quad \text{as} \quad r_k \to 0. \tag{18.2.4}$$

Proof The fact that $\nabla^2 P(x, r_k)$ is positive-definite for r_k sufficiently small means that x_k is the unique minimum of $P(x, r_k)$ as $r_k \to 0$. We now show, by contradiction, that $c_1(x_k), \ldots, c_l(x_k)$ all tend to zero as $r_k \to 0$.

Suppose this statement is false and that, for some positive constant ϵ,

$$\sum_{i=1}^{l} c_i(x_k)^2 > \epsilon \quad \text{for all } r_k.$$

Then

$$P(x_k, r_k) > F(x_k) + \frac{1}{r_k}\epsilon.$$

Now let F^* be the least value of $F(x)$ at a feasible point. Because x_k is the unique minimum of $P(x, r_k)$ it must be the case that

$$P(x_k, r_k) \leq F^*.$$

Therefore

$$F(x_k) + \frac{1}{r_k}\epsilon < F^*.$$

Rearranging, we get

$$F(x_k) < F^* - \frac{1}{r_k}\epsilon.$$

But, as $r_k \to 0$, this implies that $F(x_k)$ can be arbitrarily large and negative, which contradicts the condition that $F(x)$ is bounded below. Therefore, as $r_k \to 0$,

$$c_i(x_k) \to 0, \quad i = 1, \ldots, l. \tag{18.2.5}$$

At each unconstrained minimum, x_k,

$$\nabla P(x_k, r_k) = \nabla F(x_k) + \frac{1}{r_k}\sum_{i=1}^{l} 2c_i(x_k)\nabla c_i(x_k) = 0. \tag{18.2.6}$$

If we define

$$\tilde{\lambda}_i(x_k) = -\frac{2}{r_k}c_i(x_k) \tag{18.2.7}$$

then (18.2.6) is equivalent to

$$\nabla P(x_k, r_k) = \nabla F(x_k) - \sum_{i=1}^{l} \tilde{\lambda}_i(x_k)\nabla c_i(x_k) = 0. \tag{18.2.8}$$

Now suppose that, as $r_k \to 0$, the limit point of the sequence $\{x_k\}$ is $\bar{x}$ and that $\bar{\lambda}_i = \tilde{\lambda}_i(\bar{x})$, for $i = 1, \ldots, l$. Then, from (18.2.5) and (18.2.8),

$$c_i(\bar{x}) = 0, \quad i = 1, \ldots, l \tag{18.2.9}$$

$$\nabla F(\bar{x}) - \sum_{i=1}^{l} \bar{\lambda}_i \nabla c_i(\bar{x}) = 0. \tag{18.2.10}$$

Hence $\bar{x}$ satisfies the optimality conditions for problem (16.2.1), (16.2.2). But the assumptions imply that the problem has a unique solution x^* and unique multipliers $\lambda_1^*, \ldots, \lambda_l^*$. Therefore (18.2.3) and (18.2.4) must hold.

This result motivates the *Sequential Unconstrained Minimization Technique* (SUMT) outlined below. Propositions similar to (18.2.3) can still be proved under weaker assumptions about the problem (16.2.1), (16.2.2) and so, in practice, SUMT can usually be applied successfully without the need for a strict verification of the properties of the function and constraints. A full theoretical background to SUMT is given by Fiacco and McCormick [20].

Penalty function SUMT (P-SUMT)

Choose an initial guessed solution x_0
Choose a penalty parameter r_1 and a constant $\beta(< 1)$
Repeat for $k = 1, 2, \ldots$
starting from x_{k-1} use an iterative method to find x_k to solve (18.2.2)
set $r_{k+1} = \beta r_k$
until $||c(x_k)||$ is sufficiently small

This algorithm is an example of an *infeasible* or *exterior-point* approach. The iterates x_k do not satisfy the constraints until convergence has occurred. The method does not directly calculate the Lagrange multipliers at the solution, but we can deduce their values using (18.2.4). (It should now be clear that we have used a weak form of the penalty function approach in the formulation of some of our example problems such as R1, OC1 and OC2.)

The rate of convergence of P-SUMT can be viewed in two parts. Convergence to each penalty function minimum will be governed by the choice of unconstrained method (i.e., we can expect it to be at least superlinear if we use a quasi-Newton or Newton technique.) However, the convergence of the computed minima x_k to the solution x^* is typically linear because the errors $||x_k - x^*||$ are proportional to r_k.

An example

We can demonstrate the penalty function approach on Example (16.3.1). The penalty function associated with this problem is

$$P(x, r) = x_1^2 + 3x_1 x_2 + \frac{1}{r}(x_1 + 5x_2 - 1)^2.$$

For any value of r, the minimum of $P(x, r)$ satisfies the equations

$$\frac{\partial P}{\partial x_1} = 2x_1 + 3x_2 + \frac{2}{r}(x_1 + 5x_2 - 1) = 0$$

$$\frac{\partial P}{\partial x_2} = 3x_1 + \frac{10}{r}(x_1 + 5x_2 - 1) = 0.$$

The second equation gives $(x_1 + 5x_2 - 1) = -3rx_1/10$ and on substitution in the first equation we get $x_2 = -7x_1/15$. Eliminating x_2 from the second equation gives $(9r - 40)x_1 - 30 = 0$ and so the minimum of $P(x, r)$ is at

$$x_1 = \frac{30}{(9r - 40)}, \quad x_2 = -\frac{210}{(135r - 600)}.$$

Hence, as $r \to 0$, the minima of $P(x, r)$ tend to

$$x_1^* = -\frac{3}{4}, \quad x_2^* = \frac{7}{20}$$

which can be shown to solve (16.3.1) by direct use of the optimality conditions. The value of the constraint in (16.3.1) at the minimum of $P(x, r)$ is

$$c_1(x) = \frac{30}{(9r - 40)} - \frac{1050}{(135r - 600)} - 1 = -\frac{9r}{(9r - 40)}$$

and hence

$$-\frac{2}{r}c_1(x) = \frac{18}{(9r - 40)}.$$

If we let $r \to 0$ in the right hand side we can use (18.2.4) to deduce that the Lagrange multiplier $\lambda_1^* = -9/20$. This agrees with the result obtained directly from the optimality conditions.

Exercises

1. Use a penalty function approach to solve the problem

$$\text{Minimize} \quad x_1^3 + x_2^2 \quad \text{subject to} \quad x_2 - x_1^2 = 1.$$

2. Write down the penalty function $P(x, r)$ for Problem VLS1a and hence obtain an expression for $\hat{x}(r)$. (*Hint*: use the Sherman–Morrison–Woodbury formula (10.5.6).) Show that, as $r \to 0$, $\hat{x}(r)$ approaches the solution of the problem.

3. Suppose that we have obtained x_k, x_{k+1} as the unconstrained minima of $P(x, r_k)$ and $P(x, r_{k+1})$, respectively. Show how linear extrapolation could be used to obtain a first estimate of the minimum of $P(x, r_{k+2})$. Could we use a similar technique to predict the overall solution $x^*(= \lim_{r_k \to 0} x_k)$?

18.3.   The augmented Lagrangian

It might be imagined that we could accelerate the progress of P-SUMT by choosing r_1 to be very small in the hope of getting an acceptable estimate of x^* after only one unconstrained minimization. In practice, however, this is not a good idea because of the limitations of finite-precision arithmetic.

When r is near zero, the second term in $P(x, r)$ may dominate the first and so, when we evaluate P, the contribution of the objective function may be lost in rounding error. Numerical evaluations of ∇P and $\nabla^2 P$ are also likely to be inaccurate when r is small. In particular, $\nabla^2 P$ is said to become *ill-conditioned* when $r \to 0$ because its condition number, defined as

$$\frac{\text{maximum eigenvalue of } \nabla^2 P}{\text{minimum eigenvalue of } \nabla^2 P},$$

can get arbitrarily large. As a consequence of all this, the numerical solution of the Newton equation $(\nabla^2 P)p = -\nabla P$ is very susceptible to rounding error when $r \approx 0$ and the resulting search directions can be inaccurate and ineffective. Similar difficulties can occur during the minimization of $P(x, r)$ by quasi-Newton or conjugate gradient methods.

The only way to avoid these numerical difficulties is to ensure that values of the $c_i(x)$ are already near-zero by the time we are dealing with very small values of r. We can best achieve this if we follow the SUMT algorithm and obtain $x_1, x_2, \ldots$ by relatively easy minimizations using moderately large values of the penalty parameter so that a near-feasible approximation to x^* is available by the time the unconstrained algorithm has to deal with r close to zero.

The ill-conditioning difficulties which occur when minimizing $P(x, r)$ have motivated the use of another form of penalty function [53].

Definition The *augmented Lagrangian* is given by

$$M(x, v, r) = F(x) + \frac{1}{r} \sum_{i=1}^{l} \left(c_i(x) - \frac{r}{2} v_i \right)^2. \tag{18.3.1}$$

Compared with $P(x, r)$, the function M involves extra parameters $v_1, \ldots, v_l$ and can also be written as

$$M(x, v, r) = F(x) - \sum_{i=1}^{l} v_i c_i(x) + \frac{1}{r} \sum_{i=1}^{l} c_i(x)^2 + \frac{r}{4} \sum_{i=1}^{l} v_i^2.$$

If we assume (16.2.1), (16.2.2) has a unique solution x^* (where linear independence of $\nabla c_1(x^*), \ldots, \nabla c_l(x^*)$ implies uniqueness of the multiplier vector λ^*) then we can establish important properties of the augmented Lagrangian.

Proposition The function (18.3.1) has a stationary point at $x = x^*$ for all values of r if the parameters v_i are chosen so that $v_i = \lambda_i^*$, $i = 1, \ldots, l$.

Proof Differentiating (18.3.1) we get

$$\nabla M(x, v, r) = \nabla F(x) + \frac{1}{r} \sum_{i=1}^{l} 2 \left(c_i(x) - \frac{r}{2} v_i \right) \nabla c_i(x). \qquad (18.3.2)$$

and because $c_i(x^*) = 0$ for $i = 1, \ldots, l$, it follows that

$$\nabla M(x^*, v, r) = \nabla F(x^*) - \sum_{i=1}^{l} v_i \nabla c_i(x^*).$$

If we set $v_i = \lambda_i^*$ $(i = 1, \ldots, l)$ then condition (16.2.4) implies $\nabla M(x^*, \lambda^*, r) = 0$.

Proposition Suppose that ρ, σ are positive constants such that, when $r < \rho$ and $\|v - \lambda\| < \sigma$, the Hessian matrix $\nabla^2 M(x, v, r)$ is positive-definite for all x. Suppose also that x_k solves

$$\text{Minimize } M(x, v_k, r). \qquad (18.3.3)$$

Then, for all $r < \rho$,

$$x_k \to x^* \quad \text{as} \quad v_k \to \lambda^*. \qquad (18.3.4)$$

Moreover

$$v_{k,i} - \frac{2}{r_k} c_i(x_k) \to \lambda_i^* \quad \text{as} \quad x_k \to x^*. \qquad (18.3.5)$$

Proof The result (18.3.4) follows because we have already shown that M has a stationary point at x^* when $v = \lambda^*$. The additional conditions ensure that this stationary point is a minimum. Moreover, the relationship (18.3.5) follows because $\nabla M(x_k, v_k, r_k) = 0$ and a comparison between the terms in (18.3.2) and the corresponding ones in (16.2.4) implies the required result.

Hence we can locate x^* by minimizing M when the penalty parameter r is chosen "sufficiently small". This is not the same as requiring r to tend to zero and so it follows that we can use the penalty function (18.3.1) in a sequential unconstrained minimization technique without encountering the ill-conditioning difficulties which can occur with the function $P(x, r)$ as $r \to 0$.

A sequential unconstrained minimization approach based on the augmented Lagrangian $M(x, v, r)$ needs a method of adjusting the v

parameters so that they tend towards the Lagrange multipliers. A suitable technique is given in the following algorithm. The update that it uses for the parameter vector v_{k+1} is based on (18.3.5) which shows the relationship between the Lagrange multipliers and the constraints as $r \to 0$. As with the algorithm P-SUMT, the augmented Lagrangian approach can still be used in practice even when the strict conditions leading to (18.3.4) cannot be verified.

Augmented Lagrangian SUMT (AL-SUMT)

Choose an initial guessed solution x_0
Choose a penalty parameter r_1 and a constant $\beta(< 1)$
Choose an initial parameter vector v_1
Repeat for $k = 1, 2, \ldots$
 starting from x_{k-1} use an iterative method to find x_k to solve (18.3.3)
 set $v_{k+1} = v_k - 2c(x_k)/r_k$ and $r_{k+1} = \beta r_k$
until $\|c(x_k)\|$ is sufficiently small

Exercise
Obtain expressions for the gradient and Hessian of the augmented Lagrangian function M for the equality constrained problems VLS1a and OC1(3).

A worked example

We now demonstrate the augmented Lagrangian approach on Example (16.3.1). For this problem,

$$M(x, v, r) = x_1^2 + 3x_1x_2 - v(x_1 + 5x_2 - 1) + \frac{1}{r}(x_1 + 5x_2 - 1)^2.$$

For any value of r, the minimum of $M(x, v, r)$ satisfies the equations

$$\frac{\partial M}{\partial x_1} = 2x_1 + 3x_2 - v + \frac{2}{r}(x_1 + 5x_2 - 1) = 0 \qquad (18.3.6)$$

$$\frac{\partial M}{\partial x_2} = 3x_1 - 5v + \frac{10}{r}(x_1 + 5x_2 - 1) = 0. \qquad (18.3.7)$$

If we take $v = 0$ and $r = 0.1$ as our initial parameter choices then we can solve (18.3.6), (18.3.7) and show that the minimum of $M(x, 0, 0.1)$ occurs at

$$x_1 \approx -0.7675, \quad x_2 \approx 0.3581.$$

The value of the constraint at this point is approximately 0.023 and, by (18.3.5), the next trial value for v is

$$v \approx 0 - \frac{2}{0.1}(0.023) = -0.46.$$

With this value of v (but still with $r = 0.1$), equations (18.3.6), (18.3.7) become

$$22x_1 + 103x_2 = 20 + v = 19.54$$

$$103x_1 + 500x_2 = 100 + 5v = 97.7.$$

These yield $x_1 \approx -0.7495$ and $x_2 \approx 0.3498$ and so $c \approx -0.0005$. The new value of v is

$$v = -0.46 - \frac{2}{0.1}(-0.0005) \approx -0.45.$$

We can see that the method is giving x_1, x_2 and v as improving approximations to the solution values of (16.3.1), namely

$$x_1^* \approx -0.75, \quad x_2^* \approx 0.35, \quad \lambda^* \approx -0.45.$$

Exercises
1. Repeat the solution of the worked example above, but using -0.5 as the initial guess for the v-parameter in the augmented Lagrangian.
2. Apply the augmented Lagrangian method to the problem

Minimize $\quad F(x) = x_1^2 - 4x_1x_2 + 4x_2^2 \quad$ subject to $\quad x_1 + 3x_2 + 1 = 0.$

18.4. Results with P-SUMT and AL-SUMT

P-SUMT and AL-SUMT are OPTIMA implementations of the sequential unconstrained minimization techniques based on $P(x, r)$ and $M(x, v, r)$. The unconstrained minimizations can be done with either QNp or QNw.

Table 18.2 shows the results for Problem TD1a with the initial penalty parameter $r_1 = 0.1$ and the rate of decrease of r given by $\beta = 0.25$. For AL-SUMT the initial v-parameter vector is taken as $v_1 = 0$. For each SUMT iteration, k, Table 18.2 shows the values of the function $F(x_k)$, the constraint norm $||c(x_k)||$ and the cumulative numbers of QNw iterations and function calls used for the unconstrained minimizations so far.

We can see how successive unconstrained minima converge towards the constrained solution. Note that, for P-SUMT, the rate of reduction of the constraint norm is approximately the same as the scaling factor β. AL-SUMT, however, reduces $||c||$ more rapidly. Adjustment of

		P-SUMT			AL-SUMT	
k	$F(x_k)$	$\|c(x_k)\|$	QNw Cost	$F(x_k)$	$\|c(x_k)\|$	QNw Cost
1	35.02	5.8×10^{-2}	25/34	35.02	5.8×10^{-2}	25/34
2	35.07	1.4×10^{-2}	27/41	35.09	1.8×10^{-5}	27/41
3	35.08	3.7×10^{-3}	29/48	35.09	7.8×10^{-6}	28/46
4	35.09	9.1×10^{-4}	32/56			
5	35.09	2.3×10^{-4}	33/62			
6	35.09	5.7×10^{-5}	34/68			
7	35.09	1.4×10^{-5}	35/74			

Table 18.2. P-SUMT and AL-SUMT solutions to Problem TD1a.

the v-parameters speeds up convergence of AL-SUMT, whereas P-SUMT depends only on the reduction of r to drive the iterates x_k towards the constrained optimum.

Table 18.3 shows how performance of P-SUMT is affected by changes in the initial penalty parameter. (In each case the scaling factor is $\beta = 0.25$.) These results, for Problem TD2a, show that P-SUMT becomes appreciably less efficient as smaller values of r_1 are used. This confirms the comments made in Section 18.3, that it is better to start with a moderately large value of the penalty parameter in order to ensure that we have near-feasible and near-optimal starting points for the minimizations of $P(x, r)$ when r is very small.

	P-SUMT Iterations	QNw Cost
$r_1 = 1$	9	35/72
$r_1 = 10^{-1}$	7	32/72
$r_1 = 10^{-2}$	5	62/107
$r_1 = 10^{-3}$	4	165/308

Table 18.3. P-SUMT solutions to Problem TD2a for varying r_1.

Table 18.4 summarises the performance of P-SUMT and AL-SUMT on the test problems TD1a–OC3, showing the differences between perfect and weak line searches. The quoted figures were all obtained with the standard initial parameter settings $r_1 = 0.1$, $\beta = 0.25$ and $v_1 = 0$. For some of the examples, better results might have been obtained if we had used different values for r_1 and β: the interested reader can use the OPTIMA software to investigate this.

Table 18.4 confirms that AL-SUMT is usually more efficient than P-SUMT. However, a comparison with corresponding figures in Tables 17.1 and 18.1 shows that both the SUMT approaches often take more iterations than SOLVER for the linearly constrained problems. When the constraints are nonlinear the SUMT approaches can be more competitive (as in the case of AL-SUMT/p applied to Problems VLS2a and TLS2a).

	TD1a	TD2a	VLS1a	VLS2a	TLS1a	TLS2a
P-SUMT/w	35/74	32/72	13/36	46/91	145/231	25/47
P-SUMT/p	26/126	22/115	12/36	26/145	123/450	16/69
AL-SUMT/w	28/46	22/39	7/17	36/69	149/228	25/47
AL-SUMT/p	20/97	16/86	6/17	21/132	158/569	16/69

	OC1a(4)	OC2a(4)	FBc	OC3(6)
P-SUMT/w	37/96	49/115	54/126	88/245
P-SUMT/p	30/93	38/157	47/376	82/259
AL-SUMT/w	19/45	23/45	28/53	48/112
AL-SUMT/p	16/43	17/76	16/169	45/121

Table 18.4.   Total QN iterations/function calls for P-SUMT and AL-SUMT.

On Problems TLS1a and TLS2a, AL-SUMT does not outperform P-SUMT. This appears to be because the Lagrange multipliers for both problems are zero and so the classical penalty function – by chance – is the same as the augmented Lagrangian with $v = \lambda^*$. The calculated values of the v-parameters in AL-SUMT will be slightly worse estimates of the true Lagrange multipliers.

Exercises

1. Use P-SUMT to solve Problem TD1a and use the Lagrange multipliers to deduce an estimate of the surface area for a target volume $V^* = 21$. Check your estimate by solving a suitably modified form of Problem TD1a.

2. Obtain results like those in Tables 18.2 and 18.3 but using Problem VLS1a. Do you observe any behaviour that is different from that described in the section above? If so, can you explain why it occurs?

3. Use P-SUMT to solve Problem OC1a(4). Deduce from the Lagrange multipliers what the objective function would be if s_f were increased from 1.5 to 2. Check your predictions by solving a modified version of Problem OC1a(4).

4. Perform numerical tests for Problem TD2a to discover how the speed of convergence of AL-SUMT varies with the initial choice of r_1. Why is the performance ultimately the same as P-SUMT?

5. Perform tests to estimate the choices of r_1 and β for which AL-SUMT solves Problems VLS2a, OC3(6) in the smallest number of iterations. How do these results compare with the figures for SOLVER?

6. Using the results in Table 18.4, comment on the advantages and drawbacks of using a perfect line search in P-SUMT and AL-SUMT.

18.5.   Exact penalty functions

The approaches described so far are based on converting a constrained problem to a sequence of unconstrained ones. It is also possible to solve (16.2.1), (16.2.2) via a single unconstrained minimization. A function whose unconstrained minimum coincides with the solution to a constrained minimization problem is called an *exact penalty function*. As an example, consider

$$E(x,r) = F(x) + \frac{1}{r}\left\{\sum_{i=1}^{l}|c_i(x)|\right\}. \tag{18.5.1}$$

This is called the l_1 penalty function and it has a minimum at x^* for all r sufficiently small. It has no parameters requiring iterative adjustment and a solution of (16.2.1), (16.2.2) can be found by minimizing (18.5.1). In making this remark, of course, we assume that r has been chosen suitably. In fact there is a "threshold" condition ($r < 1/||\lambda^*||_\infty$) but normally this cannot be used in practice because the Lagrange multipliers will not be known in advance.

The function E has the undesirable property of being nonsmooth because its derivatives are discontinuous across any surface for which $c_i(x) = 0$. This fact may cause difficulties for many unconstrained minimization algorithms which assume continuity of first derivatives.

For equality constrained problems there is a smooth exact penalty function,

$$E'(x,r) = F - c^T(AA^T)^{-1}Ag + \frac{1}{r}c^Tc \tag{18.5.2}$$

where c is the vector of constraints $c_i(x)$, A is the Jacobian matrix whose rows are the constraint normals $\nabla c_i(x)^T$ and g is the gradient vector $\nabla F(x)$. The second term on the right of (18.5.2) includes a continuous approximation to the Lagrange multipliers. This follows because λ^* can be obtained from the Lagrangian stationarity condition, $g - A^T\lambda^* = 0$, by solving

$$(AA^T)\lambda^* = -Ag.$$

Hence E' is a form of augmented Lagrangian function in which the multiplier estimates, λ, vary continuously instead of being adjusted at periodic intervals. The use of (18.5.2) was first proposed by Fletcher and Lill [24] and subsequent work based on the idea is summarised in [26].

As with (18.5.1), the exact penalty function E' has a practical disadvantage. The right hand side of (18.5.2) involves first derivatives of the function and constraints and so second derivatives of F and c_i have to be obtained if E' is to be minimized by a gradient method. Worse still, third derivatives will be needed if we wish to use a Newton algorithm.

Exercise

Solve Problem TD1a by forming and minimizing the exact penalty function (18.5.2) (e.g., by using SOLVER as an unconstrained minimization method). What is the largest value of r with which you obtain the correct solution?

Investigate what happens when SOLVER is applied to the nonsmooth penalty function (18.5.1).

Chapter 19

Sequential Quadratic Programming

Sequential quadratic programming (SQP) methods have become more popular than the SUMT approaches. There have been two strands of development in this area. One involves the use of successive QP approximations to (16.2.1), (16.2.2) based on linearisations of the c_i and a quadratic model of F. Another approach uses QP subproblems which are derived from the unconstrained minimization calculations in AL-SUMT.

19.1.   Quadratic/linear models

In what follows we write

$$g(x) = \nabla F(x), \quad G(x) = \nabla^2 F(x), \quad c = (c_1, \ldots, c_m)^T. \qquad (19.1.1)$$

We also let A denote the matrix whose ith row is $\nabla c_i(x)^T$.

The first-order optimality conditions at the solution (x^*, λ^*) of the equality constrained problem (16.2.1), (16.2.2) are

$$g(x^*) - \sum_{i=1}^{l} \lambda_i^* \nabla c_i(x^*) = 0 \quad \text{and} \quad c_i(x^*) = 0, \quad i = 1, \ldots, l.$$

If x, λ are estimates of x^*, λ^*, we can introduce an error measure

$$T(x, \lambda) = \left\| g(x) - \sum_{i=1}^{l} \lambda_i \nabla c_i(x) \right\| + \kappa \| c_i(x) \| \qquad (19.1.2)$$

where κ is a positive weighting parameter.

M. Bartholomew-Biggs, *Nonlinear Optimization with Engineering Applications*,
DOI: 10.1007/978-0-387-78723-7_19, © Springer Science+Business Media, LLC 2008

Now suppose that $\delta x = x^* - x$. Then δx and λ^* satisfy

$$g(x + \delta x) - \sum_{i=1}^{l} \lambda_i^* \, \nabla c_i(x + \delta x) = 0$$

and

$$c_i(x + \delta x) = 0, \quad \text{for} \quad i = 1, \dots, l.$$

Using first order Taylor expansions we see that δx and λ^* approximately satisfy

$$g(x) + G(x)\delta x - \sum_{i=1}^{l} \lambda_i^* \{\nabla c_i(x) + \nabla^2 c_i(x)\delta x\} = 0 \qquad (19.1.3)$$

and

$$c_i(x) + \nabla c_i(x)^T \delta x = 0, \quad \text{for} \quad i = 1, \dots, l. \qquad (19.1.4)$$

If we define

$$\hat{G} = G(x) - \sum_{i=1}^{m} \lambda_i^* \, \nabla^2 c_i(x) \qquad (19.1.5)$$

then, on dropping the explicit dependence on x, (19.1.3), (19.1.4) simplify to

$$\hat{G}\delta x - A^T \lambda^* = -g \qquad (19.1.6)$$

and

$$-A\delta x = c. \qquad (19.1.7)$$

By comparing (19.1.6), (19.1.7) with (17.1.4) we see that these are optimality conditions for the quadratic programming problem

$$\text{Minimize} \quad \frac{1}{2}(\delta x^T \hat{G} \delta x) + g^T \delta x \quad \text{subject to} \quad c + A\delta x = 0. \qquad (19.1.8)$$

Hence δx and λ^* can be approximated by solving EQP (19.1.8). The objective function in (19.1.8) involves the gradient of the objective function, but its Hessian $\hat{G}$ includes second derivatives of the constraints and hence is an estimate of $\nabla^2 L$ rather than of the Hessian G. Thus nonlinearities in the constraints do appear in the problem (19.1.8), even though its constraints are only linearisations of the c_i.

The EQP (19.1.8) can be used to calculate a search direction in an iterative algorithm for a general equality constrained minimization problem. The version of the algorithm outlined below uses a quasi-Newton estimate of $\hat{G}$ rather than calculating (19.1.5) from second derivatives. We refer to this as a *Wilson–Han–Powell* algorithm because these authors (independently) did much of the pioneering work in this area [52], [64], [33].

Wilson–Han–Powell SQP algorithm (WHP-SQP)

Choose an initial point x_0 and an initial matrix B_0 approximating
(19.1.5)

Repeat for $k = 0, 1, 2 \ldots$

Obtain p_k and λ_{k+1} by solving the QP subproblem

$$\text{Minimize} \quad \frac{1}{2} p^T B_k p + \nabla F(x_k)^T p$$

$$\text{subject to} \quad c_i(x_k) + \nabla c_i(x_k)^T p = 0 \quad i = 1, \ldots, l$$

Obtain a new point $x_{k+1} = x_k + s p_k$ via a line search.
Obtain B_{k+1} by a quasi-Newton update of B_k
until $T(x_{k+1}, \lambda_{k+1})$, given by (19.1.2), is sufficiently small

The line search in WHP-SQP may be based on ensuring $P(x_{k+1}) < P(x_k)$, where P denotes some penalty function. Various choices for P have been tried. Some authors recommend the l_1 exact penalty function (18.5.1) but others use versions of the augmented Lagrangian. The line search is important because, by forcing a reduction in a composite function involving both F and the c_i, it ensures that the new point x_{k+1} is, in a measurable sense, an improvement on x_k, thereby providing a basis for a proof of convergence. The WHP-SQP approach is shown in [52] to be capable of superlinear convergence providing the updating strategy causes B_k to agree with the true Hessian of the Lagrangian in the tangent space of the constraints.

The quasi-Newton update in WHP-SQP is typically performed using the modified BFGS formula [52] based on the gradient of the Lagrangian function as outlined in section 18.1.

Exercises

1. Perform one iteration of WHP-SQP applied to the problem

$$\text{Minimize} \quad x_2 \quad \text{subject to} \quad x_1^2 + x_2^2 = 1$$

starting from $x_1 = x_2 = \frac{1}{2}$ and using $B = \nabla^2[x_2 + \lambda(x_1^2 + x_2^2 - 1)]$ with $\lambda = 1$.

2. Perform one iteration of WHP-SQP applied to

$$\text{Minimize} \quad x_2^2 \quad \text{subject to} \quad x_1^2 + x_2^2 = 1 \quad \text{and} \quad x_1 + x_2 = 0.75$$

starting from $x_1 = x_2 = \frac{1}{2}$ and using $B = \nabla^2(x_2^2)$.

19.2.   SQP methods based on penalty functions

In the Wilson–Han–Powell SQP algorithm there is no necessary connection between the QP which gives the search direction and the penalty
function used in the line search. We now derive an SQP algorithm in
which the subproblem and the step length calculation are more closely
related.   In fact, the QP subproblem approximates the minimum of
the augmented Lagrangian function (18.3.1).   A Taylor expansion of
$\nabla M(x, v, r)$ about x gives

$$\nabla M(x + \delta x, v, r) = g - A^T v + \frac{2}{r} A^T c + \left(\bar{G} + \frac{2}{r} A^T A \right) \delta x + O(\|\delta x\|^2)$$

$$(19.2.1)$$

where

$$\bar{G} = \nabla^2 F(x) - \sum_{i=1}^{l} \nabla^2 c_i(x) v_i + \frac{2}{r} \left[\sum_{i=1}^{l} \nabla^2 c_i(x) c_i(x) \right]. \qquad (19.2.2)$$

When $x = x^*$ and $v = \lambda^*$ then, because all the $c_i(x^*)$ are zero, (19.2.2)
gives

$$\bar{G} = \nabla^2 F(x^*) - \sum_{i=1}^{l} \lambda_i^* \nabla^2 c_i(x^*).$$

Hence we can regard $\bar{G}$ as an approximation to $\nabla^2 L$.

   If $x + \delta x$ minimizes $M(x, v, r)$ then the left-hand side of (19.2.1) is
zero. Hence, neglecting higher-order terms and rearranging, we get

$$\left(\bar{G} + \frac{2}{r} A^T A \right) \delta x = -g + A^T v - \frac{2}{r} A^T c. \qquad (19.2.3)$$

Solving (19.2.3) gives δx as the Newton step towards the minimum of
$M(x, v, r)$. If we now define

$$u = v - \frac{2}{r}(A\delta x + c)$$

then we can also write

$$A\delta x = -\frac{r}{2}(u - v) - c. \qquad (19.2.4)$$

Hence (19.2.3) simplifies to

$$\bar{G}\delta x - A^T u = -g. \qquad (19.2.5)$$

Comparing (19.2.5), (19.2.4) with (17.1.4) we can see that δx and u are, respectively, the solution and the Lagrange multipliers of the EQP

$$\text{Minimize} \quad \frac{1}{2}(\delta x^T \bar{G} \delta x) + g^T \delta x \quad \text{subject to} \quad c + A\delta x = -\frac{r}{2}(u - v).$$
(19.2.6)

From (19.2.4) we get

$$c(x + \delta x) \approx c + A\delta x = -\frac{r}{2}(u - v).$$

This is a first-order estimate of constraint values at the minimum of $M(x, v, r)$. If $||\delta x||$ and $||c||$ are both small (which will be the case when x is near a solution) then $u \approx v$. Hence the constraints in (19.2.6) tend to linearisations of the actual problem constraints, even when $r \neq 0$. It follows that u – the Lagrange multipliers for (19.2.6) – can also be regarded as approximating the multipliers of the original problem.

Equations (19.2.4), (19.2.5) can be rewritten as the symmetric system

$$\bar{G}\delta x - A^T u = -g$$
(19.2.7)

$$-A\delta x - \frac{r}{2}u = c - \frac{r}{2}v.$$
(19.2.8)

If we define

$$\delta v = -\frac{2}{r}(A\delta x + c),$$

so that $u = v + \delta v$ then we can rewrite (19.2.7), (19.2.8) in terms of δx and δv to obtain

$$\bar{G}\delta x - A^T \delta v = -g + A^T v$$
(19.2.9)

$$-A\delta x - \frac{r}{2}\delta v = c.$$
(19.2.10)

It can also be shown (Exercise 1, below) that we can obtain u and δx to satisfy (19.2.7), (19.2.8) by solving

$$\left(\frac{r}{2}I + A\bar{G}^{-1}A^T\right)u = A\bar{G}^{-1}g - c + \frac{r}{2}v$$
(19.2.11)

and then using

$$\delta x = \bar{G}^{-1}(A^T u - g).$$
(19.2.12)

We can now give an algorithm based on the preceding discussion. As with WHP-SQP, we describe a version which uses a quasi-Newton update. In this case we use an estimate of the matrix $\bar{G}^{-1}$ which approximates the inverse Hessian of the Lagrangian. If $H_k \approx G^{-1}$ at the start of iteration k then a search direction p_k and multipliers u_k are obtained by solving a QP subproblem of the same form as (19.2.6), namely,

$$\text{Minimize} \quad \frac{1}{2} p^T H_k^{-1} p + p^T \nabla F(x_k)$$

$$\text{subject to} \quad c(x_k) + A(x_k) p = -\frac{r_k}{2}(u_k - \lambda_k)$$

where λ_k are the Lagrange multiplier estimates at the start of the iteration. The subproblem solution is based on (19.2.11) and (19.2.12).

Augmented Lagrangian SQP algorithm (AL-SQP)

Choose initial values x_0, λ_0, r_0
Choose a matrix H_0 approximating the inverse of (19.2.2)
Choose a scaling factor $\beta < 1$. Set $\mu = 0$, $T^- = T(x, \lambda_0)$.
Repeat for $k = 0, 1, 2, \ldots$
Compute $c_k = c(x_k)$, $g_k = G(x_k)$ and $A_k = A(x_k)$.
Obtain p_k and u_k from

$$\left(\frac{r_k}{2} I + A_k H_k A_k^T \right) u_k = A_k H_k g_k - c_k + \frac{r_k}{2} \lambda_k \qquad (19.2.13)$$

$$p_k = H_k(A_k^T u_k - g_k) \qquad (19.2.14)$$

Obtain a new point $x_{k+1} = x_k + s p_k$ via a line search to give

$$M(x_{k+1}, \lambda_k, r_k) < M(x_k, \lambda_k, r_k)$$

If $T(x_k, u_k) < T^-$ then
set $r_{k+1} = \beta r_k$, $\lambda_{k+1} = u_k$ and $T^- = T(x_{k+1}, \lambda_{k+1})$
otherwise
set $r_{k+1} = r_k$ and $\lambda_{k+1} = \lambda_k$
Obtain H_{k+1} by a quasi-Newton update of H_k
until $T(x_{k+1}, \lambda_{k+1})$ is sufficiently small

The update for H_{k+1} uses the quasi-Newton condition $H_{k+1} \gamma_k = \delta_k$ in which $\gamma_k = \nabla L(x_{k+1}) - \nabla L(x_k)$ and L is the approximate Lagrangian, given by

$$L(x) = F(x) - \sum_{i=1}^{l} \lambda_{k+1} c_i(x).$$

As was discussed in relation to the reduced-gradient algorithm, it is preferable to use an updating strategy such as the modified BFGS formula [52] that ensures H_k is positive-definite.

AL-SQP can be viewed as a method for constructing an approximation to a trajectory of augmented Lagrangian minima. The parameters r and λ are adjusted as soon as a better estimate of an optimal point is found, rather than after a complete minimization of M. This gives a quicker approach to x^* than that offered by AL-SUMT.

SQP algorithms based on the penalty function $P(x, r)$ were first suggested by Murray [49] and Biggs [10]. The augmented Lagrangian version AL-SQP given above was first described in [8].

It can be shown that the subproblems in AL-SQP are guaranteed to have a solution. This is a significant advantage over the QP subproblems of WHP-SQP in which the linearisations of nonlinear constraints may be inconsistent even when the original constraints give a well-defined feasible region. An overview of developments of both the WHP-SQP and the AL-SQP approaches can be found in [26].

Exercises

1. Show that the solution of (19.2.7), (19.2.8) can be obtained by solving

$$\left(\frac{r}{2}I + A\bar{G}^{-1}A^T\right)u = A\bar{G}^{-1}g - c + \frac{r}{2}v$$

and then using

$$\delta x = \bar{G}^{-1}(A^T u - g).$$

Show also that these expressions are together algebraically equivalent to (19.2.3) and provide an alternative way of calculating the Newton step δx.

2. How could AL-SQP be modified if we wanted to get the search direction and Lagrange multiplier estimates on each iteration from (19.2.4) and (19.2.5) but using a matrix B which is a quasi-Newton estimate of the matrix $\bar{G}$ given by (19.2.2)?

3. Derive an algorithm similar to AL-SQP which is based on estimating the minimum of $P(x, r)$ rather than $M(x, v, r)$.

A worked example

Consider the problem

$$\text{Minimize } F(x) = \frac{1}{2}(x_1^2 + 2x_2^2) \quad \text{subject to } c_1(x) = x_1 + x_2 - 1 = 0.$$
$$(19.2.15)$$

Note that F is quadratic and c_1 is linear and so (19.2.2) gives $\bar{G} = \nabla^2 F(x)$. Suppose $x_1 = x_2 = 1$ is a trial solution and that $\lambda_1 = v$ is the initial Lagrange multiplier estimate. We now show the result of one iteration of AL-SQP for any v and penalty parameter r. We have $g = \nabla F = (1, 2)^T$ and $c_1 = 1$. Moreover

$$\bar{G} = \nabla^2 F = \begin{pmatrix} 1 & 0 \\ 0 & 2 \end{pmatrix} \quad \text{and so} \quad H = \bar{G}^{-1} = \begin{pmatrix} 1 & 0 \\ 0 & \frac{1}{2} \end{pmatrix}.$$

The matrix A is simply $(1, 1)$. Hence

$$AHA^T = (1, \ 1) \begin{pmatrix} 1 & 0 \\ 0 & \frac{1}{2} \end{pmatrix} \begin{pmatrix} 1 \\ 1 \end{pmatrix} = \frac{3}{2}$$

$$AHg = (1, \ 1) \begin{pmatrix} 1 & 0 \\ 0 & \frac{1}{2} \end{pmatrix} \begin{pmatrix} 1 \\ 2 \end{pmatrix} = 2.$$

Equation (19.2.13) now becomes

$$\left(\frac{r}{2} + \frac{3}{2} \right) u = 2 - 1 + \frac{r}{2} v$$

so that

$$u = \frac{2}{r + 3} \left(1 + \frac{r}{2} v \right).$$

Equation (19.2.14) then gives

$$p = \begin{pmatrix} 1 & 0 \\ 0 & \frac{1}{2} \end{pmatrix} \begin{pmatrix} u - 1 \\ u - 2 \end{pmatrix}$$

from which we get

$$p_1 = \frac{2}{r + 3} \left(1 + \frac{r}{2} v \right) - 1 = \frac{rv - r - 1}{r + 3}$$

$$p_2 = \frac{1}{r + 3} \left(1 + \frac{r}{2} v \right) - 1 = \frac{rv - 2r - 4}{2(r + 3)}.$$

Hence, the new approximation, $x + p$, to the minimum of $M(x, v, r)$ is given by

$$x^+ = \left(1 + \frac{rv - r - 1}{r + 3}, \ 1 + \frac{rv - 2r - 4}{2(r + 3)} \right)^T.$$

It is now clear that, for any value of v, x^+ tends to the solution of (19.2.15) at $(\frac{2}{3}, \frac{1}{3})^T$ as $r \to 0$.

The Lagrange multiplier at the solution of (19.2.15) can easily be shown to be $\lambda_1^* = \frac{2}{3}$. If we use the parameter value $v = \lambda_1^* = \frac{2}{3}$ in the calculations for u and p then (19.2.13) gives

$$u = \frac{2}{r + 3} \left(1 + \frac{r}{2} v \right) = \frac{2}{r + 3} \left(1 + \frac{r}{3} \right) = \frac{2}{3}.$$

Now from (19.2.14) we get

$$p_1 = -\frac{1}{3} \quad \text{and} \quad p_2 = -\frac{2}{3}.$$

Hence, when $v = \lambda_1^*$, we get $x^+ = x^*$ and $u = \lambda_1^*$ for any value of r.

Exercises

1. Repeat the worked example from this section but calculating u and p from (19.2.7) and (19.2.8).
2. Perform one iteration of AL-SQP applied to

   Minimize $F(x) = x_2^2$ subject to $x_1^2 + x_2^2 = 1$ and $x_1 + x_2 = 0.75$

   starting from $x_1 = x_2 = \frac{1}{2}$ and using $B = \nabla^2 F(x)$. How does this compare with the behaviour of WHP-SQP?

19.3.   Results with AL-SQP

AL-SQP denotes the OPTIMA implementation of the augmented Lagrangian SQP method, in which the inverse Hessian estimate H is obtained using the Powell modification to the BFGS update [52]. This method is only implemented with a weak line search.

If we apply AL-SQP to Problem TD1a (with $r_1 = 0.1$, $\beta = 0.25$) we obtain the convergence history shown in Table 19.1.

k	$F(x_k)$	$\|c(x_k)\|$	Itns/ Function Calls
1	3.646	3.2×10^{-1}	1/3
2	3.549	2.8×10^{-1}	5/11
3	3.502	5.8×10^{-1}	11/22
4	3.509	1.3×10^{-3}	12/23
5	3.509	1.1×10^{-4}	13/24

Table 19.1.   AL-SQP solution to Problem TD1a with $r_1 = 0.1$.

Table 19.1 shows progress at the end of each iteration which produces a "sufficiently large" decrease in the Kuhn–Tucker error measure T (19.1.2). On these "outer" iterations the algorithm adjusts the values of r and the multiplier estimates λ_k. By comparing Table 19.1 with Table 18.2 we see that AL-SQP converges faster than either of the SUMT methods with the same values of r_1 and β. Progress towards the solution is much more rapid when penalty parameter and multiplier estimates are updated frequently, rather than being changed only after an exact minimization of the augmented Lagrangian.

Table 19.2 shows the numbers of iterations and function calls needed by AL-SQP to solve Problems TD1a–OC3. Comparison with Table 18.4

TD1a	TD2a	VLS1a	VLS2a	TLS1a	TLS2a
14/24	7/9	4/6	27/41	57/107	23/37

OC1a(4)	OC2a(4)	OC1a(6)	OC2a(6)	FBc	OC3(6)
7/8	7/10	8/11	10/16	43/71	10/13

Table 19.2.   Performance of AL-SQP on Problems TD1a–OC3.

confirms the advantage of the SQP approach over SUMT. The figures in Tables 17.1 and 18.1 show that AL-SQP and SOLVER give broadly similar performance on nonlinearly constrained problems.

Exercise

1. Apply AL-SQP to a modified form of Problem TD1a in which the target volume is $V^* = 21$.

2. Apply AL-SQP to a modified form of Problem VLS2a which involves an extra data point $(5, 0.1)$.

3. Extend the comparison between SUMT and SQP methods to include Problems OC1a(10), OC2a(10) and OC3(10).

4. The sensitivity problem (15.4.1) can be expressed as a constrained minimization problem

   Minimize $(x_1 - x_1^*)^2 + (x_2 - x_2^*)^2$ s.t. $2x_1 x_2 + 40x_2^{-1} + 20x_1^{-1} = 35.44$.

   Solve this problem using AL-SQP.

5. Use the results in Tables 17.1, 18.1, 18.4 and 19.2 to discuss the relative performance of reduced-gradient, SUMT and SQP approaches on problems with linear and nonlinear constraints.

6. Transform Problems TD1a and TD2a using the $y_i = x_i^2$ substitution to obtain solutions which exclude negative dimensions. Solve the transformed problems using AL-SQP (or other available software) and compare the results with those for the unmodified problems.

7. Compare the Lagrange multiplier values calculated by P-SUMT, AL-SUMT and AL-SQP on the problems in Table 19.2 and comment on their accuracy.

8. Use the problems in Table 19.2 to investigate how the performance of AL-SQP can be improved by adjustment of the initial penalty parameter r_1 and the reduction factor β.

Overhead costs and runtimes

As in Chapter 14, we can use the optimal control problems to compare the numbers of iterations and function evaluations needed by constrained

	OC3(50)	OC3(75)	OC3(100)
P-SUMT	503/1336	716/1949	804/2099
AL-SUMT	297/684	420/980	535/1256
AL-SQP	48/79	62/109	70/118

Table 19.3.   Performance of SUMT and SQP on Problem OC3.

minimization problems for larger numbers of variables and constraints. Results for Problem OC3 are shown in Table 19.3.

In comparison with P-SUMT, AL-SQP uses about one-tenth as many iterations and AL-SUMT takes about three-fifths as many. The results in Table 19.3 were all obtained using weak line searches and with the standard settings $r_1 = 0.1$, $\beta = 0.25$. Obviously the behaviour would be somewhat different if other choices were made but the figures quoted give a good indication of the relative efficiencies of the SQP and SUMT approaches.

We now consider how the counts of iterations and function calls in Table 19.3 translate into execution times. We define

$$\rho_1 = \frac{\text{runtime needed by AL-SQP}}{\text{runtime needed by P-SUMT}}$$

$$\rho_2 = \frac{\text{runtime needed by AL-SUMT}}{\text{runtime needed by P-SUMT}}.$$

For OC3(50) we find $\rho_1 \approx 0.35$ and $\rho_2 \approx 0.5$. For OC3(75) we get $\rho_1 \approx 0.41$ and $\rho_2 \approx 0.47$. The relative runtimes of the two SUMT methods seem to be roughly proportional to numbers of iterations but the computational cost per iteration of AL-SQP is evidently greater than that of each QN iteration within SUMT. This can be explained by the fact that each quasi-Newton step in P-SUMT and AL-SUMT does a similar amount of work – namely, obtaining a search direction from a matrix-vector product costing n^2 multiplications. For AL-SQP, however, the calculation of a search direction involves the formation and solution of the $l \times l$ system of equations (19.2.13) which costs about $nl^2 + ln^2 + \frac{1}{6}l^3$ multiplications. Hence, if $l = qn$ (where we assume $q \leq 1$) each AL-SQP search direction is about qn times as expensive as a SUMT one.

In the case of Problem OC3, $l \approx \frac{1}{2}n$ and so the search direction calculation in AL-SQP costs about $\frac{3}{4}n^3$ multiplications. If we assume that the other operations on each iteration – the line search, the matrix update and so on – are similar for both methods then if AL-SQP takes k_1 iterations and P-SUMT takes k_2 iterations we can expect ρ_1 to satisfy a relationship of the form

$$\rho_1 \approx \frac{k_1}{k_2}(\alpha n + \beta). \tag{19.3.1}$$

The entries in Table 19.3 indicate that α and β can be obtained from

$$0.35 = \frac{48}{503}(50\alpha + \beta); \quad 0.41 = \frac{62}{716}(75\alpha + \beta).$$

and so we deduce

$$\rho_1 \approx \frac{k_1}{k_2}(0.043n + 1.53). \tag{19.3.2}$$

Thus, for large n, the time advantage of AL-SQP over P-SUMT may not be as significant as the iteration counts suggest.

For Problems OC1a and OC2a, l does not depend on n. In such cases, the cost of forming and solving (19.2.13) varies roughly with n^2 and ρ_1 can be expected to be more nearly proportional to k_1/k_2. When this happens (and also when $l = qn$ and $q \ll 1$) Al-SQP is likely to have a greater advantage over SUMT.

Exercises

1. Use the results in Table 19.3 to derive an expression similar to (19.3.2) for ρ_2, the ratio of run-times for AL-SUMT and P-SUMT on Problem OC3.

2. Obtain results similar to those in Table 19.3 to compare the performance of the SUMT and SQP approaches on Problem OC2a(n). Use measured runtimes to deduce expressions similar to (19.3.2) to give the relative runtimes of AL-SQP and AL-SUMT compared with P-SUMT.

19.4.   SQP line searches and the Maratos effect

We have already mentioned that a penalty function can be used as the basis of the line search in WHP-SQP. For problems with highly nonlinear constraints, penalty function line searches can experience a difficulty which can be explained by considering a problem with just one constraint $c(x) = 0$. Suppose x_k is an estimate of the solution and p is the search direction given by an SQP subproblem at x_k. If the constraint is nonlinear and if $c(x_k)$ is close to zero then it is possible that

$$||x_k + p - x^*|| < ||x_k - x^*|| \quad \text{and also} \quad |c(x_k + p)| > |c(x_k)|.$$

In such a case it would probably be appropriate to accept the new point $x_k + p$. However, if the line search is based on the exact penalty function

$$E(x, r) = F(x) + \frac{1}{r}|c(x)|$$

then, for small values of r, the increase in constraint violation may imply

$$E(x_k + p, r) > E(x_k, r)$$

and so the line search will reject $x_k + p$. The subsequent step-length calculation may yield $x_{k+1} = x_k + sp$ where $s \ll 1$. This phenomenon is called the *Maratos effect* [46]. It can sometimes cause very slow convergence of SQP methods when the iterates are close to the constraints (especially when near the solution).

The situation just outlined can arise whatever penalty function is used in the SQP line search. All penalty functions involve a weighted combination of the function and the constraints and the Maratos effect can occur whenever the constraint contribution is overemphasised. Unfortunately there are no hard-and-fast rules for choosing penalty parameters to ensure that the function and constraints are well balanced on every SQP iteration.

Replacing line searches with filters

We may be able to avoid the Maratos effect by dealing with function values and constraint violations separately, rather than combining them in a penalty function. We give a brief description of this approach, again using an example with a single equality constraint $c(x) = 0$.

We let x_0 be the initial guessed solution and write $F_k = F(x_k)$, $e_k = |c(x_k)|$. If p is the search direction then we can accept the new point $x_1 = x_0 + sp$ if

$$\text{either} \quad F_1 < F_0 \quad \text{or} \quad e_1 < e_0.$$

If only one of these inequalities holds then both x_0 and x_1 are included in a list of reference points called a *filter*. However, if both inequalities are satisfied then x_1 is said to *dominate* x_0 and the filter will contain only the point x_1.

Now let us suppose the filter consists of x_0 and x_1. Then the next SQP iteration will accept a point $x_2 = x_1 + sp$ if, for $j = 0, 1$,

$$\text{either} \quad F_2 < F_j \quad \text{or} \quad e_2 < e_j.$$

That is, a new point must be better (in terms of either function value or constraint violation) than all the points in the current filter. If this happens, the point x_2 will be added to the filter for use on the next iteration. Furthermore, either x_0 or x_1 can be removed from the filter if it is dominated by x_2.

We can illustrate the idea by using a plot in (F, e)-space. In Figure 19.1, A, B, C and D are points in the filter at the start of an iteration.

Any new point below or to the left of the dotted line is acceptable (e.g., P, Q and R). However, P does not dominate any of A, B, C or D

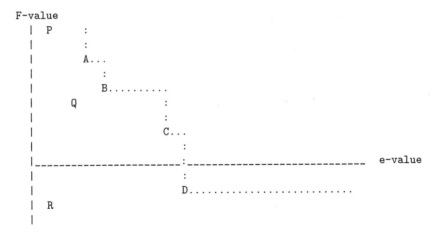

Figure 19.1.   An illustration of a filter.

and if this represents the new point then the filter for the next iteration will be defined by P, A, B, C, D. The point Q dominates A and B and if this were the outcome of the SQP step then the next filter would be Q, C, D. Finally, if R were the new point then it would dominate all of the current filter and the next iteration would only accept points to left of or below R.

The above description can be extended easily to problems with several constraints if we let e_k denote $||c(x_k)||$. For more details, see the original work by Fletcher and Leyffer [23].

Chapter 20

Inequality Constrained Optimization

20.1. Problems with inequality constraints

Up till now, we have dealt with restrictions like $x_i \geq 0$ by using transformations of the variables such as $x_i = y_i^2$. We now consider better ways of handling inequality constraints. This enables us to solve the problems listed below (some of which are extended forms of earlier examples).

Problem TD1b is a version of TD1a with lower bounds on the tank dimensions:

$$\text{Minimize} \quad 2x_1x_2 + 2x_1x_3 + x_2x_3 \tag{20.1.1}$$

$$\text{subject to} \quad x_1x_2x_3 = V^* \quad \text{and} \quad x_i \geq x_{\min}, \quad i = 1, 2, 3. \tag{20.1.2}$$

Problem TD2b is a similar variant of the maximum-volume problem TD2a.

$$\text{Minimize} \quad -x_1x_2x_3 \tag{20.1.3}$$

$$\text{s.t.} \quad 2x_1x_2 + 2x_1x_3 + x_2x_3 = S^* \quad \text{and} \quad x_{\max} \geq x_i \geq x_{\min}, \quad i = 1, 2, 3. \tag{20.1.4}$$

We can obtain modified forms of the data-fitting problems VLS1 and VLS2 by restricting the maximum size of the residuals. *Problem VLS1b* is

$$\text{Minimize} \quad \sum_{i=1}^{m}(z_i - x_1 - x_2t_i)^2 \tag{20.1.5}$$

$$\text{subject to} \quad r_{\max} \geq z_i - x_1 - x_2t_i \geq -r_{\max} \quad \text{for} \quad i = 1, \ldots, m. \tag{20.1.6}$$

M. Bartholomew-Biggs, *Nonlinear Optimization with Engineering Applications*,
DOI: 10.1007/978-0-387-78723-7_20, © Springer Science+Business Media, LLC 2008

Problem VLS2b is

$$\text{Minimize} \quad \sum_{i=1}^{m}(z_i - x_1 e^{x_2 t_i})^2 \qquad (20.1.7)$$

$$\text{subject to} \quad r_{\max} \geq z_i - x_1 e^{x_2 t_i} \geq -r_{\max} \quad \text{for} \quad i = 1, \ldots, m. \quad (20.1.8)$$

The optimal control problems from Section 3.3 can be extended to include bounds on the size of the applied accelerations. *Problem OC1b(n)* is

$$\text{Minimize} \quad x_1^2 + x_n^2 + \sum_{k=2}^{n}(x_k - x_{k-1})^2 \qquad (20.1.9)$$

s.t. $s_n = s_f$ and $u_n = u_f$ and $x_{\max} \geq x_i \geq -x_{\max}$ for $i = 1, \ldots, n$
$$(20.1.10)$$

where s_n and u_n are given by (3.3.1).

Problem OC2b(n) has the same constraints (20.1.10) but a different objective function

$$\text{Minimize} \quad x_1^2 + x_n^2 + \sum_{k=2}^{n}\left(1 - \frac{x_k}{x_{k-1}}\right)^2. \qquad (20.1.11)$$

Problem OC3a(n) is an inequality constrained variant of Problem OC3 given by (16.1.12)–(16.1.16) which features an upper bound on speed. It is written as

$$\text{Minimize} \quad x_1^2 + x_n^2 + \sum_{i=2}^{n}(x_i - x_{i-1})^2 \qquad (20.1.12)$$

subject to

$$u_k - u_{k-1} - (x_k - c_D u_k^2)\tau = 0 \quad \text{for} \quad k = 1, \ldots, n-1 \quad (20.1.13)$$

$$u_{\max} \geq u_k \quad \text{for} \quad k = 1, \ldots, n-1 \qquad (20.1.14)$$

$$u_f - u_{n-1} - (x_n - c_D u_f^2)\tau = 0 \qquad (20.1.15)$$

$$s_n - s_f = 0 \qquad (20.1.16)$$

where s_n is given by

$$s_k = s_{k-1} + u_{k-1}\tau + \frac{1}{2}(x_k - c_D u_k^2)\tau^2 \quad \text{for} \quad k = 1, \ldots, n. \quad (20.1.17)$$

We can also consider a constrained version of the preventive maintenance problem (see Section 13.2) in which the variables $x_1, \ldots, x_n$ (the intervals between PMs) are subject to a lower bound. From the values of these x_i we deduce other dependent variables

$$t_n = \sum_{k=1}^{n} x_k; \quad y_k = \left(\sum_{j=1}^{k-1} b_j x_j\right) + x_k \quad \text{and} \quad y_{k-1}^+ = y_{k-1} + (1 - b_{k-1})x_{k-1}.$$

Using notation from the section preceding (13.2.3), *Problem PM1a(n)* is

$$\text{Minimize} \quad \frac{\gamma_r + (n-1) + \gamma_m\{\hat{H}(y_1) + \sum_{k=2}^{n}[\hat{H}(y_k) - \hat{H}(y_{k-1}^+)]\}}{t_n}$$

$$(20.1.18)$$

$$\text{subject to} \quad x_i \geq x_{\min}, \quad \text{for} \quad i = 1, \ldots, n \qquad (20.1.19)$$

where γ_r, γ_m and $\hat{H}(t)$ are given in (13.2.4).

Before considering optimality conditions for inequality constrained problems we first consider some new optimization applications.

Minimax approximation

One way of fitting a model $z = \phi(x, t)$ to a dataset (t_i, z_i), $i = 1, \ldots, m$ is to find values of the parameters x_i so as to minimize the sum of squared residuals

$$\sum_{i=1}^{m} (z_i - \phi(x, t_i))^2.$$

An alternative approach would be to choose the x_i to minimize the largest residual. This is called the *minimax* problem which is posed as

$$\text{Minimize} \quad \max_{1 \leq i \leq m} |z_i - \phi(x, t_i)|. \qquad (20.1.20)$$

Clearly the objective function in (20.1.20) is nonsmooth and is therefore more difficult to minimize than a sum of squared terms. However, we can also calculate minimax approximations by solving a differentiable constrained minimization problem. If there are n parameters x_i appearing in the model function ϕ then a solution to (20.1.20) can be obtained from *Problem MMX(n)*

$$\text{Minimize} \quad x_{n+1} \qquad (20.1.21)$$

$$\text{subject to} \quad x_{n+1} \geq z_i - \phi(x, t_i) \geq -x_{n+1} \quad i = 1, \ldots, m. \qquad (20.1.22)$$

Hence, if we were fitting the model $z = \phi(x, t) = x_1 t$ to the data

$$t_1 = 1, \ z_1 = 1; \quad t_2 = 2, \ z_2 = 1.5$$

the minimax solution would be obtained from

$$\text{Minimize} \quad x_2$$

$$\text{subject to} \quad x_2 \geq 1 - x_1 \geq -x_2; \quad x_2 \geq 1.5 - 2x_1 \geq -x_2.$$

It is interesting to compare minimax approximation with least squares. We consider an example involving data from a test on the reliability of a certain electrical component. For a sample batch of components we record the fraction, z, still surviving in working order after t months of continuous operation. The sample originally contained 1000 items and the number surviving is shown in Table 20.1

Months	0	1	2	3	4	5	6
Survivors	1	0.947	0.894	0.848	0.792	0.740	0.693

Months	7	8	9	10	11	12
Survivors	0.656	0.610	0.572	0.535	0.518	0.514

Table 20.1. Monthly data for component failures.

Suppose we seek to fit a straight line to this data. Solving the minimax problem (20.1.21), (20.1.22) (by methods described in later chapters) the best approximation is found to be

$$z = 0.967 - 0.0405t.$$

If we fit a straight line to the same data using the least squares approach then the best approximation is

$$z = 0.973 - 0.0427t.$$

Although these two lines are similar, they do represent different ways of fitting the data. This can be seen in Figure 20.1. The least-squares line stays closer to the majority of the data points whereas the minimax line sacrifices this closeness in order to reduce the error at the last point on the graph. This point $(12, 0.514)$ is the one which deviates most from the line of the rest of the data. It is quite common for minimax approximations to pay more attention to such points which are sometimes called *outliers*.

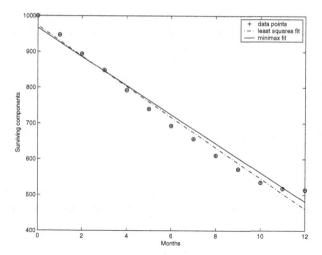

Figure 20.1. Comparing minimax and least squares fits to failure data.

Worst-case optimization

The minimax approach to data-fitting is also closely related to the idea of *worst-case analysis*. If we want to optimize a complex enginee-ring system we may choose to do so in a way which gives the "best" performance under "normal" operating conditions. Another approach, however, would be to configure the system so as to counteract the adverse consequences of abnormal behaviour. To put it another way, we could devise an optimal business plan which seeks to maximize manufacturing profit under the assumption that customer demand and raw material supply remain more or less constant or we could put the emphasis on minimizing the losses that would occur if some foreseeable (but fairly unlikely) worst-case event were to occur such as a major change in currency exchange rates that damaged export prospects.

This can be illustrated if we consider the optimal control problems OC1 and OC2. In these problems we seek to optimize a function which represents an overall measure of the smoothness of the train's motion. An alternative would be to seek to minimize the worst instance of non-smoothness. Thus, instead of minimizing the sum of squared terms

$$\sum_{i=1}^{n}(x_i - x_{i-1})^2$$

we could minimize

$$\max_{1 \leq i \leq m} |x_i - x_{i-1}|.$$

Introducing an extra variable as in the previous section, we can use notation from the relevant Section of chapter 3 and pose *Problem OC4(n)* as

$$\text{Minimize} \quad \rho(x_{n+1} + x_1^2 + x_n^2) + (s - s_f)^2 + (u - u_f)^2$$
$$(20.1.23)$$

$$\text{subject to} \quad -x_{n+1} \le (x_i - x_{i-1}) \le x_{n+1} \quad \text{for} \quad i = 2, \ldots, n.$$
$$(20.1.24)$$

This is an inequality constrained quadratic programming problem. We can also consider *Problem OC5(n)* which has nonlinear constraints and is

$$\text{Minimize} \quad \rho(x_{n+1}^2 + x_1^2 + x_n^2) + (s - s_f)^2 + (u - u_f)^2 \quad (20.1.25)$$

$$\text{subject to} \quad (x_i - x_{i-1})^2 \le x_{n+1}^2 \quad \text{for} \quad i = 2, \ldots, n. \quad (20.1.26)$$

20.2.   Optimality conditions

The problems we have discussed in the previous section are all instances of the general nonlinear programming problem of finding $x_1, \ldots, x_n$ to solve

$$\text{Minimize} \quad F(x) \qquad (20.2.1)$$

$$\text{subject to} \quad c_i(x) = 0, \quad i = 1, \ldots, l \qquad (20.2.2)$$

$$\text{and} \quad c_i(x) \ge 0, \quad i = l+1, \ldots, m. \qquad (20.2.3)$$

There is no loss of generality in writing constraints in the form (20.2.3) because an inequality such as $x_1 + x_2 \le 1$ can also be expressed as $1 - x_1 - x_2 \ge 0$.

Definition If F and all the c_i in (20.2.1)–(20.2.3) are linear functions then this is a *linear programming* (LP) problem. Specialised solution methods for this case can be found in [63].

Definition If, in (20.2.1)–(20.2.3), the function F is quadratic and the c_i are linear then it is a *quadratic programming* (QP) problem.

Definition If x satisfies the equality and inequality constraints (20.2.2), (20.2.3) it is said to be *feasible*. Otherwise it is called *infeasible*.

First-order optimality conditions at a solution of (20.2.1)–(20.2.3) are extensions of the KKT conditions already stated for equality constrained problems.

Proposition If x^* is a local solution to (20.2.1)–(20.2.3) then the optimality conditions are as follows. The solution x^* must be feasible, and so

$$c_i(x^*) = 0, \quad i = 1, \ldots, l \tag{20.2.4}$$

$$c_i(x^*) \geq 0, \quad i = l+1, \ldots, m. \tag{20.2.5}$$

Furthermore, the Lagrange multipliers λ_i^*, $i = 1, \ldots, m$ associated with the constraints must satisfy

$$\nabla L(x^*, \lambda^*) = \nabla F(x^*) - \sum_{i=1}^{m} \lambda_i^* \, \nabla c_i(x^*) = 0 \tag{20.2.6}$$

$$\lambda_i^* c_i(x^*) = 0, \quad i = l+1, \ldots, m \tag{20.2.7}$$

$$\text{and} \quad \lambda_i^* \geq 0, \quad i = l+1, \ldots, m. \tag{20.2.8}$$

The so-called *complementarity condition* (20.2.7) states that an inequality constraint is either satisfied as an equality at x^* or it has a zero Lagrange multiplier.

Definition If $l+1 \leq i \leq m$ and $c_i(x^*) = 0$ we say that the ith inequality is *binding*, and x^* lies on an *edge* of the feasible region.

Definition If $\lambda_i^* = 0$ when $l+1 \leq i \leq m$ then the ith inequality is said to be *nonbinding* and x^* is *inside* the ith constraint boundary.

Binding and nonbinding constraints are illustrated in Figure 20.2 which shows the contours of an objective function and three linear inequality constraints which define the feasible region as the interior and edges of the triangle ABC. The unconstrained minimum is at (0, 0) in the centre of the figure and hence the constrained minimum is at the point X lying on the edge AC. The constraint represented by AC is binding but those represented by AB and BC are nonbinding.

The nonnegativity condition (20.2.8) on the Lagrange multipliers for the inequality constraints ensures that the function F will not be reduced by a move off any of the binding constraints at x^* to the interior of the feasible region.

The uniqueness of the Lagrange multipliers depends on the normals to the binding constraints at x^* being linearly independent.

Second-order optimality conditions for (20.2.1)–(20.2.3) involve feasible directions for the binding constraints at x^*. Let I^* be the set of indices

$$I^* = \{ i \mid l+1 \leq i \leq m \ \text{and} \ c_i(x^*) = 0 \} \tag{20.2.9}$$

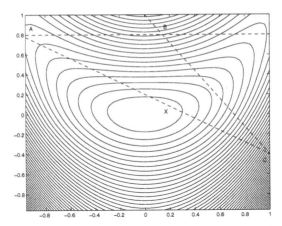

Figure 20.2.   Binding and nonbinding constraints.

and let N be the matrix whose first l rows are $\nabla c_1(x^*)^T, \dots, \nabla c_l(x^*)^T$ and whose remaining rows are $\nabla c_i(x^*)^T$ for $i \in I^*$. Then a second-order condition for x^* to be a solution of (20.2.1)–(20.2.3) is

$$z^T \nabla^2 L(x^*, \lambda^*) \geq 0 \quad \text{for any } z \text{ such that } Nz = 0. \qquad (20.2.10)$$

This is equivalent to $z^T \nabla^2 F(x^*)z \geq 0$ if all the constraints are linear.

Exercise
Consider the problem

$$\text{Minimize} \quad F(x) \quad \text{subject to} \quad c_1(x) \geq 0.$$

Suppose x^* and λ_1^* satisfy optimality condition (20.2.7) and that $c_1(x^*) = 0$ but that $\lambda_1^* < 0$. Show there is a feasible point $\tilde{x} = x^* + \delta$ for which $F(\tilde{x}) < F(x^*)$. What does this imply about the optimality of x^*?

A worked example

Consider the problem

$$\text{Minimize } F(x) = x_1^2 + 3x_2^2 \qquad (20.2.11)$$

subject to $c_1(x) = x_1 + 5x_2 - 1 \geq 0$ and $c_2(x) = x_1 + 1 \geq 0$.

$$\qquad (20.2.12)$$

The solution values x_1^*, x_2^*, λ_1^*, λ_2^* must satisfy

$$x_1 + 5x_2 - 1 \geq 0 \quad \text{and} \quad x_1 + 1 \geq 0 \qquad (20.2.13)$$

$$2x_1 - \lambda_1 - \lambda_2 = 0 \quad \text{and} \quad 6x_2 - 5\lambda_1 = 0 \qquad (20.2.14)$$

$$\lambda_1(x_1 + 5x_2 - 1) = 0 \quad \text{and} \quad \lambda_2(x_1 + 1) = 0 \qquad (20.2.15)$$

$$\lambda_1 \geq 0 \quad \text{and} \quad \lambda_2 \geq 0. \qquad (20.2.16)$$

Rather than attempting to solve this system of equations and inequalities we simply use it to test the optimality of some candidate solutions.

Suppose first that we try the point $x_1 = -1$, $x_2 = 2/5$ at which both constraints are binding. From (20.2.14), the corresponding Lagrange multipliers are

$$\lambda_1 = \frac{12}{25} \quad \text{and} \quad \lambda_2 = -\frac{62}{25}.$$

This violates (20.2.16) so we deduce that $(-1, \ 2/5)$ is not a solution.

Next we investigate the possibility of a solution at $x_1 = -1$, $x_2 = 1$ with the second constraint binding but not the first. This implies $\lambda_1 = 0$. But if we put $\lambda_1 = 0$ in the first equation of (20.2.14) then we get $\lambda_2 = 2x_1 = -2$ which violates (20.2.16). Alternatively, if we consider the second equation in (20.2.14) it implies that $\lambda_1 = 6/5$ which conflicts with the fact that λ_1 must be zero. Thus the optimality tests fail on two counts and so the feasible point $(-1, 1)$ is not a solution.

Finally we consider whether there is a solution with the first constraint binding but not the second. This will mean that $\lambda_2 = 0$ and then (20.2.14) implies

$$2x_1 - \lambda_1 = 0 \quad \text{and} \quad 6x_2 - 5\lambda_1 = 0.$$

For these equations to be consistent we need $2x_1 = 6x_2/5$; and combining this with the first equation in (20.2.13) we get $x_1 = 3/28$, $x_2 = 5/28$. It then follows that $\lambda_1 = 6/28$, $\lambda_2 = 0$ and all the first-order optimality conditions are satisfied.

Exercises

1. By considering the first-order optimality conditions for the problem

$$\text{Minimize} \quad -3x_1 - 4x_2 \quad \text{s.t.} \quad x_1 \geq 0; \quad x_2 \geq 0; \quad 1 - x_1^2 - x_2^2 \geq 0$$

determine which (if any) of the following points is a solution:
(i) $x_1 = x_2 = 0$; (ii) $x_1 = 1$, $x_2 = 0$; (iii) $x_1 = 4/5$, $x_2 = 3/5$

2. Finding the model $\phi = x_1 t$ which gives a minimax fit to the two data points $(t_i, \ z_i) = (1, 1), \ (2, 1.5)$ leads to the problem

$$\text{Minimize} \quad x_2 \quad \text{s.t.} \quad x_2 \geq 1 - x_1 \geq -x_2; \quad x_2 \geq 1.5 - 2x_1 \geq -x_2.$$

Use a rough sketch to show that the solution line must pass below the first point and above the second. This suggests that the two binding constraints are

$$x_2 - 1 + x_1 \geq 0 \quad \text{and} \quad 1.5 - 2x_1 + x_2 \geq 0.$$

Hence deduce values for the two nonzero Lagrange multipliers at the solution and check whether all the optimality conditions are satisfied.

20.3. Transforming inequalities to equalities

Before considering methods which handle inequality constraints directly, we mention briefly an approach which allows Problems (20.2.1)–(20.2.3) to be tackled by the techniques we have described in the previous three chapters. A problem with inequality constraints can be transformed into one with only equalities if we introduce extra variables. Thus the problem

$$\text{Minimize} \quad F(x) \quad \text{subject to} \quad c_i(x) \geq 0, \quad i = 1, \ldots, m \quad (20.3.1)$$

can be rewritten as

$$\text{Minimize} \quad F(x) \quad \text{subject to} \quad c_i(x) - w_i^2 = 0, \quad i = 1, \ldots, m. \quad (20.3.2)$$

Here $w_1, \ldots, w_m$ are called *squared slack variables*. The conversion of (20.3.1) into (20.3.2) can have certain benefits when both n, the number of variables, and m, the number of constraints, are quite small. However, for larger problems, the fact that (20.3.2) involves $n + m$ variables is usually a disadvantage.

Exercises

1. Solve the problem

$$\text{Minimize} \quad x_1^2 + x_2 \quad \text{subject to} \quad x_1 + x_2 \geq 3$$

by using a squared slack variable to transform it into an equality constrained problem. Show that the solution to the transformed problem satisfies the optimality conditions for the original one.

2. Show that, in general, the optimality conditions for (20.3.1) are consistent with those for (20.3.2).

3. Write down an optimization problem which uses squared slack variables to transform VLS1b into one involving just equality constraints. What is the relationship between the Lagrange multipliers for the two problems?

20.4. Transforming inequalities to simple bounds

The simplest forms of inequality constraint are simple bounds on the variables. A problem with general inequality constraints can be transformed into one with only equality constraints and simple bounds. Thus the problem (20.3.1) can be rewritten as

$$\text{Minimize} \quad F(x) \tag{20.4.1}$$

subject to $\ c_i(x) - w_i = 0, \ i = 1, \ldots, m \ $ and $\ w_i \geq 0, \quad i = 1, \ldots, m.$
$$\tag{20.4.2}$$

The quantities $w_1, \ldots, w_m$ in (20.4.2) are called *slack variables*. The conversion of (20.3.1) into (20.4.1), (20.4.2) has the advantage that the inequalities are simple enough to be handled efficiently by the reduced-gradient approach, as outlined in the next chapter. However, when m is large, (20.4.1), (20.4.2) has the disadvantage that it involves many more variables than (20.3.1).

Exercises

1. Solve the problem

$$\text{Minimize} \quad x_1^2 + x_2 \quad \text{subject to} \quad x_1 + x_2 > 3$$

using a slack variable to convert the inequality into an equation and a bound.

2. Show that the optimality conditions for (20.4.1), (20.4.2) are equivalent to those for (20.3.1).

20.5. Example problems

We now introduce some specific examples involving inequality constraints.

Problem TD1b is the tank design problem with simple bounds, given by (20.1.1), (20.1.2) with $V^* = 20$ and $x_{\min} = 1.9$. The starting guess is $x_1 = x_2 = 2$, $x_3 = 5$. The solution of the original unconstrained problem TD1 is not a feasible point for TD1b and so the lower bound on x_1 is binding. The tank dimensions are

$$x_1 = 1.9, \ x_2 = x_3 \approx 3.34 \quad \text{giving surface area } S^* \approx 35.185.$$

Problem TD2b comes from (20.1.3), (20.1.4) with $x_{\max} = 3.2$, $x_{\min} = 1.9$ and $S^* = 35$. The starting guess is $x_1 = x_2 = 2$, $x_3 = 5$ and the solution is

$$x_1 \approx 1.934, \ x_2 = x_3 = 3.2.$$

Hence the upper bounds on x_2 and x_3 are binding.

Problem VLS1b is given by (20.1.5), (20.1.6) with data points

$$(t_i, \ z_i) = (0,3), \ (1, \ 8), (2, \ 12), \ (3, \ 17)$$

and $r_{max} = 0.275$. We use the same starting guess as for the unconstrained problem VLS1. The solution is $x_1 = 3.175$, $x_2 = 4.55$. The
binding constraints involve the residuals at the second and third data
points. The optimum value of the sum of squared errors is 0.2125, which
is greater than the minimum value 0.2 which can be obtained if the constraints are removed.

Problem VLS2b is given by (20.1.7), (20.1.8) with the data points

$$(t_i, \ z_i) = (0,1), \ (1, \ 0.5), (2, \ 0.4), \ (3, \ 0.3), \ (4, \ 0.2)$$

and $r_{max} = 0.08$. We use the same starting guess as for the equality
constrained problem VLS2a. However the solution is different because of
the inequality constraints. The solution of VLS2b is at $x_1 \approx 0.924$, $x_2 \approx$
-0.466. The binding inequalities are the upper bound on the residual at
the second data point and the lower bound on the residual at the fourth
point. The minimum value of the sum of squared residuals ≈ 0.0218.
This is of course larger than the sum of squared residuals (0.0185) at
the solution of VLS2a when there are no constraints.

Problems OC1b(n) and OC2b(n) are given by (20.1.9)–(20.1.11). We use
the values

$$\tau = \frac{3}{n}, \quad u_0 = u_f = 0, \quad s_0 = 0 \quad \text{and} \quad s_f = 1.5.$$

The limiting value for the accelerations, x_i, is $x_{min} = 0.7$. The starting
guess is the same as for the corresponding equality constrained problems
OC1a(n) and OC2a(n). We can tell from the solutions to the unbounded
problem OC1a that this limit on the acceleration will influence the solution.

Problem OC3a(n) is defined by (20.1.12)–(20.1.17) using the same values
of τ, u_0, u_f and s_f as in OC1b(n) and OC2b(n), and taking $u_{max} = 0.8$.

Problem PM1a(n) involves (20.1.18), (20.1.19) with data from Section 13.2
and x_{min} taken as 0.5. The starting guess is $x_i = 0.6$, $(i = 1, \ldots, n)$.
Solutions for various values of n are the same as those given in Section
13.2 which were obtained by unconstrained minimization of the function
(13.2.3) with the transformation of variables $y_i = x_i^2$.

Problem MMX1 involves fitting a minimax straight line to the dataset
$(t_i, \ z_i)$ given by Table 20.1. This is done by solving problem (20.1.21),

(20.1.22) with $n = 3$ and $\phi(x, t_i) = x_2 + x_1 t_i$. The starting guess is $x_1 = x_2 = x_3 = 0$. The solution has $x_1 \approx 0.967$, $x_2 \approx 0.0405$, $x_3 \approx 0.033$.

Consideration of the graphs in Figure 20.1 shows that a straight line is not a particularly good choice of model function for the data in Table 20.1. The data points follow a curve which appears to be flattening out as t increases and therefore a negative exponential model might be more appropriate. This is done via *Problem MMX2* which solves (20.1.21), (20.1.22) with $n = 3$ and

$$\phi(x, t_i) = x_2 e^{x_1 t_i}.$$

The starting guess is $x_1 = x_2 = x_3 = 0$ and the solution is

$$x_1 \approx 0.984, \quad x_2 \approx -0.0574, \quad x_3 \approx 0.01954.$$

Problem OC4(n) has function and constraints given by (20.1.23) and (20.1.24). The starting guesses for the accelerations $x_1, \ldots, x_n$ are the same as for Problem OC1; and the initial value of $x_{n+1} = 1.5$. When $n = 6$ the solution is

$$x_1 = x_3 \approx 0.398, \quad x_2 = -x_5 \approx 1.19, \quad x_4 = x_6 \approx -0.393.$$

These cause the difference $x_i - x_{i-1}$ to have constant magnitude ≈ 0.792 for $i = 1, \ldots, 6$. The related *Problem OC5(n)* has function and constraints given by (20.1.25) and (20.1.26). The starting guess is the same as that for OC4(n).

Exercises

1. Formulate an inequality constrained problem based on (15.4.1) in which the objective is to determine the smallest change to the tank dimensions such that the surface area exceeds 35.5. Also derive a problem which seeks the largest change in the variables such that the surface area does not exceed 35.5.

2. In (15.4.1) and the previous question the smallest and largest changes in the variables are expressed in least-squares form. Formulate corresponding optimization problems to find the smallest (or largest) change in either x_1 or x_2 for which the surface area exceeds (does not exceed) 35.5.

Preventive maintenance

Regular oiling
of all moving parts is like
turning the clock back

which begs the question
can routine lubrication
keep the clock ticking?

Chapter 21

Extending Equality Constraint Methods to Inequalities

21.1. Quadratic programming with inequalities

When (20.2.1)–(20.2.3) is an inequality constrained quadratic programming problem (IQP), a solution procedure can be based on repeated use of the optimality conditions for an EQP. We describe this approach for an IQP which only has inequality constraints; the extension to a mixed equality-inequality problem is straightforward. We consider the problem

$$\text{Minimize } \frac{1}{2}(x^T G x) + h^T x + c \qquad (21.1.1)$$

$$\text{subject to} \quad \hat{A}x + \hat{b} \geq 0 \qquad (21.1.2)$$

and we assume first that G is positive-definite.

We begin by identifying an *active set* of constraints. This is an estimate of the ones which are binding at the solution. We might, for example, simply guess that the first t rows of $\hat{A}$ and $\hat{b}$ correspond to the active constraints (although we can usually make more informed choices than this).

We now let A be the matrix and b the vector formed from the rows of $\hat{A}$ and $\hat{b}$ corresponding to active constraints. If we treat the active constraints as if they were equalities and ignore all other constraints then we can obtain a trial solution $(\tilde{x}, \tilde{\lambda})$ by minimizing (21.1.1) subject to

$$Ax + b = 0.$$

We can do this by solving the EQP optimality conditions (17.1.4). If we find

$$\hat{A}\tilde{x} + \hat{b} \geq 0,$$

M. Bartholomew-Biggs, *Nonlinear Optimization with Engineering Applications*,
DOI: 10.1007/978-0-387-78723-7_21, © Springer Science+Business Media, LLC 2008

(so that $\tilde{x}$ does not violate any inactive constraints) and if

$$\tilde{\lambda}^T(\hat{A}\tilde{x} + \hat{b}) = 0 \quad \text{and} \quad \tilde{\lambda} \geq 0$$

then optimality conditions (20.2.4)–(20.2.8) are all satisfied and the IQP is solved by $x^* = \tilde{x}$, $\lambda^* = \tilde{\lambda}$. If, however, $\tilde{x}$ and $\tilde{\lambda}$ are not optimal we must change the active set and solve another EQP. This process can be repeated until the active set becomes the binding set for problem (21.1.1)–(21.1.2).

The choice of a new active set can be based on two considerations. The first is that any new constraints which are violated at $\tilde{x}$ can be regarded as candidates for being added to the current active set. Secondly, any active constraints which correspond to a negative element in $\tilde{\lambda}$ are candidates for deletion from the current active set.

The algorithm given below formalises the ideas just outlined. For convenience, we use $\hat{a}_i$ to denote the ith row of $\hat{A}$.

Inequality QP algorithm for a positive-definite Hessian

Choose an initial point x and set $\lambda_1 = \cdots = \lambda_m = 0$.
Repeat
Identify the *active* constraints as being those for which

$$\hat{a}_i^T x + \hat{b}_i < 0 \quad \text{or} \quad (\hat{a}_i^T x + \hat{b}_i = 0 \quad \text{and} \quad \lambda_i \geq 0)$$

Renumber constraints so the active set is $i = 1, \ldots, t$
Set $g = Gx + h$ and $b_i = \hat{a}_i^T x + \hat{b}_i$ for $i = 1, \ldots, t$
Find p and μ to solve the EQP

$$\text{Minimize} \quad \frac{1}{2}(p^T Gp) + g^T p \quad \text{s.t.} \quad \hat{a}_i^T p + b_i = 0, \quad i = 1, \ldots, t \quad (21.1.3)$$

Set $s = 1$, $\lambda_i = \mu_i$ ($i = 1, \ldots, t$) and $\lambda_i = 0$ ($i = t+1, \ldots, m$)
Repeat for $i = t+1, \ldots, m$ (i.e. for all *inactive* constraints)

$$\text{if } \hat{a}_i^T p < 0 \quad \text{set } s = \min(s, -\frac{(\hat{a}_i^T x + \hat{b}_i)}{\hat{a}_i^T p} \quad (21.1.4)$$

Replace x by $x + sp$
until the optimality conditions

$$\hat{A}x + \hat{b} \geq 0 \quad Gx + h - \hat{A}^T \lambda = 0 \quad \lambda \geq 0$$

are all satisfied.

Implementations of this approach may differ in the method of solving (21.1.3) and also in rules about how many constraints may be added to or dropped from the active set in a single iteration. The stepsize calculation (21.1.4) checks all the inactive constraints that might be violated by a step along p and ensures that no more than one constraint can be added to the active set on the current iteration. It may also be advisable to allow only one constraint at a time to be deleted from the active set although the algorithm outlined above does not impose such a restriction. (For a fuller discussion see Fletcher [25].)

We now consider the possibility that G in (21.1.1) may be indefinite. This complicates the active set approach because, even when the original problem (21.1.1), (21.1.2) has a unique solution, it may happen that the EQP (21.1.3) cannot be solved because G is not positive-definite on the subspace of feasible directions for some choices of the active set. One way to deal with this difficulty is to use the reduced-gradient method for solving the EQP subproblem (see Section 17.3). If $Z^T G Z$ in (17.3.6) is not positive-definite then negative diagonal terms will appear during an attempt to calculate its Cholesky factors. As explained in Chapter 9, there are variants of the Cholesky method (e.g., [60]) which correct such negative terms and hence implicitly create a modified reduced-Hessian which is then used to give a descent direction for the objective function. Further details of this are outside the scope of this section.

An example

Consider the IQP

$$\text{Minimize} \quad F(x) = x_1^2 + 3x_2^2$$

$$\text{subject to} \quad c_1(x) = x_1 + 5x_2 - 1 \geq 0 \quad \text{and} \quad c_2(x) = x_2 \geq 0.$$

In the notation of the general problem (21.1.1), (21.1.2),

$$G = \begin{pmatrix} 2 & 0 \\ 0 & 6 \end{pmatrix}, \quad h = \begin{pmatrix} 0 \\ 0 \end{pmatrix}, \quad c = 0, \quad \hat{A} = \begin{pmatrix} 1 & 5 \\ 0 & 1 \end{pmatrix}, \quad \hat{b} = \begin{pmatrix} -1 \\ 0 \end{pmatrix}.$$

If we choose $x = (0,0)^T$ as the starting point then both constraints are treated as active and so $A = \hat{A}$ and $b = \hat{b}$. Using the notation in the IQP algorithm,

$$g = Gx + h = \begin{pmatrix} 0 \\ 0 \end{pmatrix}$$

and we obtain a search direction p by solving the EQP subproblem

$$\text{Minimize} \quad p_1^2 + 3p_2^2$$

$$\text{subject to} \quad p_1 + 5p_2 - 1 = 0 \quad \text{and} \quad p_2 = 0.$$

Here p is determined entirely by the constraints as $(1,0)^T$ and so the new solution estimate is $x + p = (1,0)^T$. We get the Lagrange multipliers for the subproblem by solving the optimality conditions

$$2p_1 - \lambda_1 = 0; \quad 6p_2 - 5\lambda_1 - \lambda_2 = 0$$

which gives

$$\lambda_1 = 2, \quad \lambda_2 = -5\lambda_1 = -10.$$

Hence the first EQP subproblem has yielded a point which is feasible but is not optimal because λ_2 is negative.

For the next iteration we drop the second constraint from the active set and so A is just the first row of $\hat{A}$ and $b = \hat{A}x + \hat{b}_1 = 0$. Now $g = Gx + h = (2,0)^T$ and $b = 0$ and the EQP subproblem is

$$\text{Minimize} \ \ p_1^2 + 3p_2^2 + 2p_1$$

$$\text{subject to} \ \ p_1 + 5p_2 = 0.$$

Now p_1, p_2 and the Lagrange multiplier λ_1 must satisfy

$$p_1 + 5p_2 = 0; \quad 2p_1 + 2 - \lambda_1 = 0; \quad 6p_2 - 5\lambda_1 = 0.$$

Solving these equations by any method gives

$$p_1 = -\frac{25}{28}, \quad p_2 = \frac{5}{28} \ \ \text{and} \ \ \lambda_1 = \frac{6}{28}.$$

Hence after the second iteration the estimated solution $x + p$ is

$$x_1 = \frac{3}{28}, \quad x_2 = \frac{5}{28},$$

with Lagrange multipliers

$$\lambda_1 = \frac{6}{28} \ \ \text{and} \ \ \lambda_2 = 0.$$

It is left to the reader to show that this satisfies all the optimality conditions for the original problem.

Exercise
Use the IQP algorithm to solve the example problem above starting from the feasible initial guess $x = (1,1)^T$. Repeat the solution starting from the infeasible point $x = (2.5, \ -0.2)^T$.

21.2. Reduced-gradients for inequality constraints

We can combine an active-set strategy with the reduced-gradient approach from Section 17.3 to solve problems with a nonquadratic objective function $F(x)$ and linear inequality constraints $\hat{A}x + \hat{b} \geq 0$. Whenever the active set changes, however, it will necessary to recompute the Y and Z basis matrices used in (17.3.5)–(17.3.7).

Reduced-gradient algorithm for linear inequality constraints

Choose an initial feasible point x_0 and set $\lambda_0 = 0$
Choose B_0 as a positive-definite estimate of $\nabla^2 F(x_0)$.
Repeat for $k = 0, 1, 2, ..$
Set $g_k = \nabla F(x_k)$
Select active constraints as those with $c_i(x_k) = 0$ and $\lambda_{k_i} \geq 0$
Get A_k as the matrix of active constraint normals at x_k
Obtain Y_k and Z_k as basis matrices for the range and null spaces of A_k
Determine z from $Z_k^T B_k Z_k z = -Z_k^T g_k$ and set $p_k = Z_k z$
Find λ_{k+1} by solving $Y_k^T A_k^T \lambda = Y_k^T g_k + Y_k^T B_k p_k$
Perform a line search to get $x_{k+1} = x_k + s p_k$ so $F(x_{k+1}) < F(x_k)$.
Do a quasi-Newton update of B_k with $\delta = x_{k+1} - x_k$, $\gamma = g_{k+1} - g_k$
until $\|Z^T g_k\|$ is less than a specified tolerance.

Note that the calculation of the line search step in this algorithm must ensure that no new constraints are violated. Hence the stepsize s is subject to an upper limit which allows at most one new constraint to become binding at x_{k+1}. This can be calculated as in (21.1.4) in the IQP algorithm in the previous section. This means the line search will in general be weak rather than perfect.

The algorithm given above can be extended to deal with nonlinear inequalities if a restoration step is included, as described in Section 18.1.

Reduced-gradient methods for simple bounds

One case where Z is easy to calculate (which makes the reduced-gradient approach very attractive) is when a constrained optimization problem involves no equalities and all the inequalities are simple bounds on the variables

$$l_i \leq x_i \leq u_i \quad i = 1, \ldots, n.$$

In this situation we can split the variables at the start of each iteration into those which are "fixed" (i.e., on their bounds) and those which are "free". If x_k is the solution estimate at the start of iteration k

then the bound on the ith variable is active if x_{k_i} is fixed, which means that

$$(x_{k_i} = l_i \text{ and } g_{k_i} > 0) \quad \text{or} \quad (x_{k_i} = u_i \text{ and } g_{k_i} < 0). \tag{21.2.1}$$

The Z matrix whose columns span the space of the free variables can then be taken simply as a partition of the identity matrix.

When taking a step from x_k to x_{k+1} along a search direction p_k, the stepsize must ensure that no new bounds are violated. A maximum stepsize to force each free variable x_{k_i} to stay within its bounds can be calculated as

$$\sigma_i = \begin{cases} (u_i - x_{k_i})/p_{k_i} & \text{if } p_{k_i} > 0 \\ (l_i - x_{k_i})/p_{k_i} & \text{if } p_{k_i} < 0. \end{cases} \tag{21.2.2}$$

Reduced-gradient algorithm for simple bounds

Choose an initial feasible point x_0
Choose B_0 as a positive definite estimate of $\nabla^2 F(x_0)$.
Repeat for $k = 0, 1, 2, \ldots$
Set $g_k = \nabla F(x_k)$
Set Z_k to be the $n \times n$ identity matrix
Repeat for $i = 1, \ldots, n$
If x_{k_i} is such that (21.2.1) holds delete ith column of Z_k.
Solve $Z_k^T B_k Z_k z = -Z_k^T g_k$ and set $p_k = Z_k z$
Use a line search to find s so that $F(x_k + s p_k) < F(x_k)$
Repeat for each free variable x_{k_i}
  calculate σ_i from (21.2.2) and set $s = \min(s, \sigma_i)$
Set $x_{k+1} = x_k + s p_k$
Update B_k using $\delta = x_{k+1} - x_k$ and $\gamma = g_{k+1} - g_k$
until $\|Z_k^T g_k\| <$ specified tolerance.

Exercise
Consider the problem

$$\text{Minimize} \quad F(y) = y^T Q y + 100(e^T y - 1)^2$$

$$\text{subject to} \quad y_i \geq 0, \quad i = 1, \ldots, n$$

where $n = 3$, $e = (1, 1, 1)^T$ and

$$Q = \begin{pmatrix} 0.0181 & -0.0281 & -0.00194 \\ -0.0281 & 0.0514 & 0.00528 \\ -0.00194 & 0.00528 & 0.0147 \end{pmatrix}.$$

Use the reduced gradient approach to solve this problem, starting from the guess $y_1 = y_2 = 0$, $y_3 = 1$ and taking B as the true Hessian $\nabla^2 F$.

Numerical results with SOLVER

The GRG reduced-gradient method [43] implemented in SOLVER [29, 48] can be applied to inequality constrained QPs and also to more general problems with inequality constraints. As mentioned before, SOLVER is more flexible than the reduced-gradient algorithm given above and it can be started at an infeasible point. The first few iterations locate a feasible point and then the algorithm proceeds as outlined in the previous section. SOLVER can also deal with nonlinear inequality constraints by means of a restoration step strategy of the kind described in section 18.1.

Table 21.1 shows the number of iterations needed by SOLVER to converge for some of the problems listed at the end of the previous chapter. (The bracketed figures show how many iterations are needed to give a feasible point.)

TD1b	TD2b	VLS1b	VLS2b	OC1b(6)	OC2b(6)	OC3a(6)
8(2)	4(2)	4(2)	6(4)	7(2)	7(2)	15(1)

PM1a(15)	MMX1	MMX2	OC4(6)	OC5(6)
14(0)	10(3)	12(2)	16(0)	16(0)

Table 21.1.   Iteration counts for SOLVER on Problems TD1b–OC5.

Exercises

1. Obtain the Lagrange multipliers at the SOLVER solution of Problem TD1b and hence estimate the minimum surface area if the lower bound x_{min} is increased to 2.
2. Transform Problem TD2b into equality-constrained form using the squared slack variables and then apply SOLVER to find a solution.
3. Explain what happens when SOLVER is applied to VLS2b with r_{max} increased to 0.1. What happens if r_{max} is reduced to 0.07?
4. For the problems included in Table 21.1 determine how many SOLVER iterations are needed before the binding set of constraints has been identified.

21.3.   Penalty functions for inequality constraints

The P-SUMT approach can be applied to inequality constrained optimization problems if we use a modified form of the penalty function (18.2.1).

Definition A version of penalty function (18.2.1) for (20.2.1)–(20.2.3) is

$$P(x, r) = F(x) + \frac{1}{r} \left\{ \sum_{i=1}^{l} c_i(x)^2 + \sum_{i=l+1}^{m} \min[0, c_i(x)]^2 \right\}. \qquad (21.3.1)$$

The first penalty term treats the equalities in the same way as in (18.2.1) but the second involves only the violated inequalities. We can use (21.3.1) in an algorithm which is very similar to P-SUMT and which does not need to identify active constraints on every iteration. This algorithm is based on the following result.

Proposition Suppose (20.2.1)–(20.2.3) has a unique solution x^*, λ^* and that $F(x)$ is bounded below for all x. Suppose also that ρ is a positive constant and that, for all $r_k < \rho$, the Hessian matrix $\nabla^2 P(x, r_k)$ of (21.3.1) is positive-definite for all x. If x_k denotes the solution of the unconstrained problem

$$\text{Minimize } P(x, r_k) \qquad (21.3.2)$$

then $x_k \to x^*$ as $r_k \to 0$. Furthermore, if $c_i(x_k) \leq 0$ as $r_k \to 0$,

$$\lambda_i^* = \lim_{r_k \to 0} \left\{ -\frac{2}{r_k} c_i(x_k) \right\}. \qquad (21.3.3)$$

If $c_i(x_k) > 0$ as $r_k \to 0$ then $\lambda_i^* = 0$.

In essentials, the proof of (21.3.3) is similar to that for (18.2.3), (18.2.4) for equality constraints. For more details and stronger results see [20].

A worked example

Consider the problem

$$\text{Minimize } x_1^2 + 2x_2^2 \quad \text{subject to} \quad x_1 + x_2 \geq 2.$$

The penalty function for this problem is

$$P(x, r) = x_1^2 + 2x_2^2 + \frac{1}{r}[\min(0, x_1 + x_2 - 2)]^2.$$

We need to consider whether $P(x, r)$ can have a minimum at a feasible point. If this is the case then the penalty term is zero and we must have

$$\frac{\partial P}{\partial x_1} = 2x_1 = 0 \quad \text{and} \quad \frac{\partial P}{\partial x_2} = 4x_2 = 0.$$

But this implies $x_1 = x_2 = 0$ which is *not* a feasible point. This contradiction means that the minimum of P must be at an infeasible point and

$$\frac{\partial P}{\partial x_1} = 2x_1 + \frac{2}{r}(x_1 + x_2 - 2) = 0 \quad \text{and} \quad \frac{\partial P}{\partial x_2} = 4x_2 + \frac{2}{r}(x_1 + x_2 - 2) = 0.$$

This implies that $x_1 = 2x_2$ and therefore

$$4x_2 + \frac{2}{r}(3x_2 - 2) = 0.$$

and hence

$$x_2 = \frac{2}{2r + 3}, \quad x_1 = \frac{4}{2r + 3}.$$

In the limit, as $r \to 0$, the solution of the original problem is at $x = \left(\frac{4}{3}, \frac{2}{3}\right)$.

Exercises

1. Write down expressions for the gradient and Hessian of the function (21.3.1). Explain why (21.3.1) may be harder to minimize than the penalty function (18.2.1) for equality constraints only, even when F is quadratic and all the c_i are linear.

2. Use the penalty function (21.3.1) to solve the problem

$$\text{Minimize} \quad x_1^2 + x_2^2 - 2x_1 + 1 \quad \text{subject to} \quad x_1 \geq 2.$$

How would the solution change if the constraint were $x_1 \geq 0$?

21.4.   AL-SUMT **for inequality constraints**

The penalty function (21.3.1) can be difficult to minimize when r is very small for the reasons discussed in Section 18.3. However, the augmented Lagrangian function can be extended to deal with inequality constraints and leads to a SUMT approach which does not require the penalty parameter to tend to zero.

Definition   The augmented Lagrangian function $M(x, v, r)$ for use with inequality constraints has the following form [59]

$$F(x) + \frac{1}{r}\left\{\sum_{i=1}^{l}\left(c_i(x) - \frac{r}{2}v_i\right)^2 + \sum_{i=l+1}^{m}\left[\min\left(0, c_i(x) - \frac{r}{2}v_i\right)\right]^2\right\}.$$

$$(21.4.1)$$

This function has a stationary point at x^* and it can be used in the AL-SUMT approach. There are only two alterations to the algorithm in

Section 18.3. The initial choices of the parameters $v_{l+1}, \ldots, v_m$ (i.e., those for the inequality constraints) must be non-negative and the rule for updating $v_{k,i}$ $(i = l+1, \ldots, m)$ is

$$v_{k+1,i} = \begin{cases} v_{k,i} - \frac{2}{r_k} c_i(x_k) & \text{if } c_i(x_k) < \frac{r_k}{2} v_{k,i} \\ 0 & \text{otherwise} \end{cases} . \tag{21.4.2}$$

Formula (21.4.2) will cause $v_k \to \lambda^*$ as $c(x_k) \to 0$. The justification for these changes is fairly straightforward [59].

Exercises

1. Apply AL-SUMT to the problem

   $$\text{Minimize } x_1^2 + 2x_2^2 \quad \text{subject to } x_1 + x_2 \geq 2$$

   using $v = 1$ as the initial guess for the multiplier parameter.

2. There is an exact penalty function for the inequality constrained problem similar to the one given in Chapter 18 for equality constrained problems. It is

   $$E(x,r) = F(x) + \frac{1}{r} \left\{ \sum_{i=1}^{l} |c_i(x)| + \sum_{i=l+1}^{m} |\min(0, c_i(x))| \right\}. \tag{21.4.3}$$

   Use this function to solve the problem in the previous question.

21.5. SQP for inequality constraints

The ideas in Chapter 19 can be extended to deal with (20.2.1)–(20.2.3) simply by including the inequality constraints (in linearised form) in the QP subproblem. Thus, in WHP-SQP, the calculation which gives the search direction and new trial Lagrange multipliers is

Obtain p_k and λ_{k+1} by solving the QP subproblem

$$\text{Minimize } \frac{1}{2} p^T B_k p + \nabla F(x_k)^T p$$

$$\text{subject to } c_i(x_k) + \nabla c_i(x_k)^T p = 0, \quad i = 1, \ldots, l$$

$$c_i(x_k) + \nabla c_i(x_k)^T p \geq 0, \quad i = l+1, \ldots, m.$$

The rest of the algorithm WHP-SQP is essentially the same as in Section 19.1. The line search along p_k will involve a function such as (21.4.1) or (21.4.3) which handles inequality constraints.

Similarly, in AL-SQP, p_k and u_k are obtained by solving the QP subproblem

$$\text{Minimize} \quad \frac{1}{2}p^T H_k^{-1} p + p^T \nabla F(x)$$

$$\text{subject to} \quad c_i(x_k) + \nabla c_i(x_k)^T p = -\frac{r_k}{2}(u_{k_i} - \lambda_{k_i}), \quad i = 1, \dots, l$$

$$\text{and} \quad c_i(x_k) + \nabla c_i(x_k)^T p \geq -\frac{r_k}{2}(u_{k_i} - \lambda_{k_i}), \quad i = l+1, \dots, m$$

where H_k is an estimate of the inverse Hessian of the Lagrangian. This subproblem approximates the Newton direction towards the minimum of the augmented Lagrangian (21.4.1) (see [8]). The line search in the inequality-constraint version of AL-SQP is also based on obtaining a reduction in (21.4.1).

The IQP subproblems in both algorithms can be solved by an active-set approach as outlined in Section 21.1. This strategy is simplified if we ensure that the Hessian approximations B_k or H_k remain positive-definite throughout.

21.6. Results with P-SUMT, AL-SUMT and AL-SQP

Table 21.2 shows the numbers of iterations and function values needed by the OPTIMA implementations of P-SUMT, AL-SUMT and AL-SQP on test problems from Section 20.5.

Method	TD1b	TD2b	VLS1b	VLS2b	OC1b(6)	OC2b(6)	OC3a(6)
P-SUMT	46/112	48/134	26/74	21/50	104/247	106/281	106/273
AL-SUMT	33/62	33/67	11/25	10/18	63/141	63/147	68/158
AL-SQP	14/27	8/8	6/6	4/4	12/13	11/13	11/11

Method	PM1a(15)	MMX1	MMX2	OC4(6)	OC5(6)
P-SUMT	95/283	74/261	74/250	59/160	51/177
AL-SUMT	56/139	52/152	55/165	50/128	45/145
AL-SQP	13/15	5/5	7/7	17/26	20/27

Table 21.2. Performance of SUMT and SQP on Problems TD1b–OC5.

By comparing Table 21.2 with Table 21.1, we see that AL-SQP and SOLVER give comparable performance on many of the problems. AL-SQP also does much better than either of the SUMT methods in terms of the numbers of iterations and function calls needed for convergence.

When dealing with inequality constrained problems, the fact that AL-SQP uses fewer iterations than the SUMT methods must be interpreted with some caution. Each iteration of a SUMT method merely does the work of computing a search direction and updating an inverse-Hessian estimate (and this is true whether the problem has equality or inequality constraints). When SUMT uses quasi-Newton minimizations, each search direction costs only a matrix-vector multiplication. An iteration of AL-SQP, however, may be much more expensive than this. An SQP search direction is obtained by solving an IQP subproblem which may, in turn, require a number of EQP solutions if an active-set approach is used. For instance, if an iteration starts from a feasible point and if the IQP subproblem has t binding constraints then at least t EQP problems will be solved. Each one of these subproblems involves the solution of a system of equations of the form (17.1.4).

In practice, the early iterations of AL-SQP are usually more expensive than the later ones. Once the iterates are fairly close to x^*, the IQP subproblem will have a "warm start" with the initial active set being the same (or nearly the same) as the binding set and only one or two EQP steps will be needed. Even then, however, an iteration of AL-SQP may involve more linear algebra calculations than an iteration of P-SUMT or AL-SUMT. Hence the runtimes for an AL-SQP solution may not be so much less than those for an AL-SUMT solution as might be suggested by the counts of iterations and function calls in Table 21.2.

Exercises
1. Construct a table similar to 18.2 and 19.1 to compare the progress of SUMT and SQP approaches on Problems TD1b and TD2b.
2. The figures in Table 21.2 were obtained using weak line searches in the unconstrained minimizations. Construct a similar table for the case when P-SUMT and AL-SUMT use QNp rather than QNw.
3. Repeat the tests in Table 21.2 using both low- and high-accuracy convergence tests and comment on the results. Can you draw any conclusions about whether the ultimate rates of convergence of the methods are linear or superlinear?
4. Re-write TD1b and TD2b as equality-constrained problems using squared slack variables. Solve these problems using SUMT and SQP approaches and discuss the results.
5. Use the SUMT and SQP methods to solve modified versions of TD1b and TD2b in which the bounds on the tank dimensions are $x_{max} = 3.6$, $x_{min} = 1.8$.
6. Use the SUMT and SQP methods to solve modified versions of VLS1b in which r_{max} is increased to 0.28. What happens is r_{max} is reduced to 0.27?
7. Extend the comparison in Table 21.2 to include the problems OC4(n) and OC5(n) for $n = 10, 20, \ldots, 50$. If possible, compare the runtimes

for these solutions and use them as the basis for a discussion similar
to that in Section 19.3.

8. Implement a spreadsheet version of the penalty function SUMT
   approach for inequality constraints, using SOLVER as the
   unconstrained minimizer and test its performance on Problems TD1b
   and VLS2b.

9. Compare the results from the previous question with those from a
   similar spreadsheet implementation of augmented Lagrangian SUMT.

The British aircraft industry circa 1966 (Part 1) [4]

Donald's involved in a government contract
about slender deltas and laminar flow;
he's busy with transforms and multiple integrals –
equations and formulae row upon row.

Gerald is dozing and dreaming of Wimbledon
(his sister gets tickets from someone at work);
he's meant to be checking some data with Ronald
who spots a mistake, wakes him up with a jerk.

Chapter 22

Barrier Function Methods

22.1.   Problems with inequality constraints only

Throughout this chapter, we consider the following problem in which all
the constraints are inequalities:

$$\text{Minimize } F(x) \tag{22.1.1}$$

$$\text{subject to} \quad c_i(x) \geq 0, \quad i = 1, \ldots, m. \tag{22.1.2}$$

When applied to problems of this form, penalty function methods usually
produce a sequence of points $\{x_k\}$ which lie outside the region defined
by the inequalities and only approach the boundary as the iterations
converge. By contrast, barrier function methods – which can only be
applied to problems of the form (22.1.1), (22.1.2) – generate points inside
the feasible region.

   Among the example problems we have considered, those which feature
only inequality constraints are VLS1b,VLS2b, PM1a, MMX1, MMX2, OC4 and
OC5. To these we add one further example, based on a problem given
by Hersom [34] which involves optimizing the cost of operating a cutting
tool.

The machine tool problem

A machine tool consists of a cutting wheel which rotates with speed v
at the circumference. The workpiece moves past the wheel with speed
u. The motor driving the wheel also acts through a gearbox to move the
workpiece and so $u = \phi v$ where ϕ is the gear-ratio. The depth of cut
is d. Operating constraints include a limit on gear-ratio ϕ of the form
$\phi_{max}v \geq u \geq \phi_{min}v$. Other conditions are as follows:

M. Bartholomew-Biggs, *Nonlinear Optimization with Engineering Applications*,
DOI: 10.1007/978-0-387-78723-7_22, © Springer Science+Business Media, LLC 2008

Limits on motor power $\qquad\qquad v^{0.2}u^{0.8}d^{0.8}v \le P_{\max}$
Limit on shear stress in drive shaft $\qquad uv \le S_{\max}$
Bounds on cutting speed and depth $\qquad v_{\max} \ge v \ge 0; \quad d_{\max} \ge d \ge 0.$

The operating cost is made up of fixed components (such as labour costs) plus the replacement cost of the cutting wheel, C_r. If C_f denotes fixed hourly costs and if T_L is the lifetime of the wheel then the lifetime cost is $C_f T_L + C_r$. We can also say that the amount of material removed from the workpiece during the life of the wheel is proportional to $vudT_L$. The performance measure that we want to minimize is

$$\frac{\text{Total operating cost}}{\text{Total material removed}}.$$

If the life of the wheel is estimated by an expression of the form

$$T_L = \frac{\kappa}{v^{1.3}u^{1.7}d^{0.6}} \tag{22.1.3}$$

then the performance function becomes

$$\frac{1}{vudT_L}(C_f T_L + C_r) = \frac{1}{vud}\left(C_f + \frac{C_r}{T_L}\right) = \frac{1}{vud}\left(C_f + \frac{C_r}{\kappa}(d^{0.6}u^{1.7}v^{1.3})\right).$$

To give some numerical values to the parameters in this problem we suppose the cutting wheel has radius $1/6$ metres so that a rotational speed of R r.p.m. corresponds to a value $v \approx R/60$ metres/sec at the circumference. We let the rotational speed be limited by $v_{\max} = 10$ metres/sec. If the workpiece maximum speed $u_{\max} = 2.5$ cm/sec and the maximum depth of cut is $d_{\max} = 2$ cm then a value $\kappa = 0.1$ in (22.1.3) implies that the cutting wheel has a life of about 8 hours when v, u and d are at their maximum values.

The values in the previous paragraph are used in *Problem MT1* in which we let $x_1 = v$ (metres/sec), $x_2 = u$ (cm/sec) and $x_3 = d$ (cm). (This choice of units causes the values of the variables to be of broadly similar magnitudes. It is worth paying attention to the relative sizes of variables because the numerical solution of badly-scaled problems can sometimes be difficult.)

If C_{rf} denotes the ratio C_r/C_f then the problem is

$$\text{Minimize} \quad \frac{1}{x_1 x_2 x_3}(1 + C_{rf}x_1^{1.3}x_2^{1.7}x_3^{0.6}) \tag{22.1.4}$$

subject to

$$10 \ge x_1 \ge 0; \quad 2 \ge x_3 \ge 0; \quad 0.25x_1 \ge x_2 \ge 0.1x_1 \tag{22.1.5}$$

$$x_1^{0.2} x_2^{0.8} x_3^{0.8} \le P_{\max}; \quad x_1 x_2 \le S_{\max}. \tag{22.1.6}$$

In (22.1.6) $P_{\max}$ and $S_{\max}$ are figures for power and stress limits. If we take $C_{rf} = 10$, $P_{\max} = 1.5$ and $S_{\max} = 5$ then the solution to (22.1.4)–(22.1.6) has $x_1 \approx 7.1$, $x_2 \approx 0.71$, $x_3 \approx 1.44$ giving a function value of 0.15.

Exercises

1. Use SOLVER to compute solutions of problem (22.1.4)–(22.1.6) for values of C_{rf} in the range $10 \le C_{rf} \le 500$ and comment on the way the results change as the fixed costs increase relative to the tool replacement costs.

2. Plot contours and constraints of a reduced version of the machine tool problem in which d is fixed as 1.

22.2. Barrier functions

Definition One form of barrier function for the problem (22.1.1), (22.1.2) is

$$B(x, \ r) = F(x) + r \sum_{i=1}^{m} \frac{1}{c_i(x)}. \tag{22.2.1}$$

Because the barrier term includes reciprocals of the constraints, B will be much greater than F when x is a feasible point near an edge of the feasible region, causing some of the $c_i(x)$ to be near zero. On the other hand, $B \approx F$ when x is inside the feasible region and all the $c_i(x)$ are much greater than zero.

Definition A more widely used barrier function for (22.1.1), (22.1.2) is

$$B(x, \ r) = F(x) - r \sum_{i=1}^{m} \log(c_i(x)). \tag{22.2.2}$$

When $1 > c_i(x) > 0$ then $\log(c_i(x)) < 0$. Hence the second term on the right of (22.2.2) implies $B \gg F$ when any of the constraint functions is small and positive. Note, however, that (22.2.2) is undefined when any $c_i(x)$ are negative.

There is a relationship, similar to that for penalty functions, between unconstrained minima of $B(x, r)$ and the solution of (22.1.1), (22.1.2).

Proposition Suppose that (22.1.1), (22.1.2) has a unique solution x^*, λ^*. Suppose also that ρ is a positive constant and, for all $r_k < \rho$, the Hessian matrix $\nabla^2 B(x, r_k)$ of the barrier functions (22.2.1) or (22.2.2) is positive-definite for all feasible x. If x_k denotes the solution of the unconstrained problem

$$\text{Minimize } B(x, r_k) \qquad\qquad (22.2.3)$$

then $x_k \to x^*$ as $r_k \to 0$. Moreover,

$$\frac{r_k}{c_i(x_k)^2} \to \lambda_i^* \quad \text{as } r_k \to 0 \quad \text{if } B \text{ is defined by (22.2.1)} \qquad (22.2.4)$$

$$\frac{r_k}{c_i(x_k)} \to \lambda_i^* \quad \text{as } r_k \to 0 \quad \text{if } B \text{ is defined by (22.2.2)}. \qquad (22.2.5)$$

We omit the main part of the proof of this result. However it is easy to justify (22.2.4) because differentiating (22.2.1) gives

$$\nabla B(x_k, r_k) = \nabla F(x_k) - \sum_{i=1}^{m} \frac{r_k}{c_i(x_k)^2} \nabla c_i(x_k) = 0. \qquad (22.2.6)$$

By comparing (22.2.6) with the Lagrangian stationarity condition (16.2.4) as $r_k \to 0$ we deduce (22.2.4). A similar argument justifies (22.2.5).

This proposition is the basis of the B-SUMT algorithm, stated below. (A fuller theoretical background can be found in [20].) B-SUMT can often be used successfully, in practice, for problems of the form (22.1.1), (22.1.2) even when the conditions in the proposition cannot be verified.

Barrier function SUMT (B-SUMT)

Choose an initial guessed solution x_0
Choose a penalty parameter r_1 and a constant $\beta(< 1)$
Repeat for $k = 1, 2, \ldots$
starting from x_{k-1} use an iterative method to find x_k to solve (22.2.3)
set $r_{k+1} = \beta r_k$
if B is defined by (22.2.1) then

$$\hat{\lambda}_i = \frac{r_k}{c_i(x_k)^2} \quad \text{for } i = 1, \ldots, m$$

else, if B is defined by (22.2.2) then

$$\hat{\lambda}_i = \frac{r_k}{c_i(x_k)} \quad \text{for } i = 1, \ldots, m$$

until $\hat{\lambda}_1 c_1(x_k), \ldots, \hat{\lambda}_m c_m(x_k)$ are all sufficiently small.

The convergence test for the algorithm is based on satisfying the complementarity condition (20.2.7), using the estimated Lagrange multipliers implied by (22.2.4) or (22.2.5).

Exercises

1. Obtain expressions for the Hessian matrices of barrier functions
(22.2.1) and (22.2.2). Hence find an expression for the Newton search
direction for (22.2.2).   How could this expression be modified if
$(\nabla^2 F(x))^{-1}$ were available?

2. Discuss other possible stopping rules for the B-SUMT algorithm.

An example

As an example of the use of the log-barrier function we consider the
problem

$$\text{Minimize } F(x) = x_1^2 + 3x_2^2 \tag{22.2.7}$$

$$\text{subject to } c_1(x) = x_1 + 5x_2 - 1 \geq 0. \tag{22.2.8}$$

The corresponding barrier function is

$$B(x, r) = x_1^2 + 3x_2^2 - r \log(x_1 + 5x_2 - 1)$$

and hence the minimum of $B(x, r)$ satisfies

$$\frac{\partial B}{\partial x_1} = 2x_1 - \frac{r}{(x_1 + 5x_2 - 1)} = 0 \tag{22.2.9}$$

$$\frac{\partial B}{\partial x_2} = 6x_2 - \frac{5r}{(x_1 + 5x_2 - 1)} = 0. \tag{22.2.10}$$

Eliminating the term involving r between these two equations we get

$$x_2 = \frac{5}{3}x_1. \tag{22.2.11}$$

Substitution in (22.2.9) then gives

$$2x_1 \left(x_1 + \frac{25}{3}x_1 - 1 \right) - r = 0.$$

This simplifies to

$$\frac{56}{3}x_1^2 - 2x_1 - r = 0 \tag{22.2.12}$$

so that

$$x_1 = \frac{3}{112} \left(2 \pm \sqrt{4 + \frac{224r}{3}} \right).$$

Using (22.2.11) we get

$$x_2 = \frac{5}{112}\left(2 \pm \sqrt{4 + \frac{224r}{3}}\right).$$

In the expressions for x_1 and x_2, the quantity under the square root is greater than 4 when $r > 0$. Hence (22.2.12) gives one positive and one negative value for x_1. But (22.2.11) means that x_2 must have the same sign as x_1. However, a solution with both x_1 and x_2 negative cannot satisfy (22.2.8). Therefore the feasible unconstrained minimum of $B(x,r)$ is at

$$x_1 = \frac{3}{112}\left(2 + \sqrt{4 + \frac{224r}{3}}\right), \quad x_2 = \frac{5}{112}\left(2 + \sqrt{4 + \frac{224r}{3}}\right).$$

Hence, as $r \to 0$ we have $x_1 \to 3/28$ and $x_2 \to 5/28$. The reader can verify that these values satisfy the optimality conditions for problem (22.2.7), (22.2.8).

Exercises

1. Deduce the Lagrange multiplier for the worked example above.
2. Use a log-barrier function approach to solve the problem

$$\text{Minimize} \quad x_1 + x_2 \quad \text{subject to} \quad x_1^2 + x_2^2 \leq 2.$$

3. A log-barrier approach is used to solve the problem

$$\text{Minimize} \quad -c^T y \quad \text{subject to } y^T Q y \leq V_a.$$

   Suppose that the barrier parameter r is chosen so the minimum of $B(y,r)$ occurs where $y^T Q y = kV_a$, where $k < 1$. Obtain an expression for $y(r)$ which minimizes the barrier function and hence find r in terms of c, Q, and V_a.
4. Solve

$$\text{Minimize} \quad x_1 + 2x_2 \quad \text{subject to} \quad x_1 \geq 0, \ x_2 \geq 1$$

   using the barrier function (22.2.1).

22.3.   Results with B-SUMT

B-SUMT is the OPTIMA implementation of the barrier SUMT algorithm using the log-barrier function (22.2.2). In B-SUMT the unconstrained minimizations are done by QNw or QNp.

A safeguard is needed in the line search for the unconstrained minimization technique in B-SUMT. The log-barrier function is undefined if any of the constraints $c_i(x)$ are nonpositive and therefore the linesearch must reject trial points where this occurs. This can be done within an Armijo line-search by setting $B(x, r)$ to a very large value at any point x which is infeasible.

B-SUMT must be started with a feasible point and so we may have to use different initial guesses from those in the problem definitions in Section 20.5. For some problems, it is relatively easy to obtain a feasible point by inspection (e.g., when the constraints are simple bounds on the variables). It is also quite straightforward to choose a feasible point for problem OC4 by making all the variables $x_1, \ldots, x_n$ equal (so that all the constraint functions are zero) or by using the standard values of $x_1, \ldots, x_n$ and simply choosing x_{n+1} large enough to ensure that the inequalities are all satisfied. For other problems, such as VLS1b and VLS2b, it is not at all easy to pick feasible values of x_1 and x_2. In such cases we may have to use a more general approach based on solving the unconstrained problem

$$\text{Minimize} \quad F(x) = \sum_{i=1}^{m} \{\min[0, \ c_i(x)]\}^2. \qquad (22.3.1)$$

This will have an optimum value of zero at any point which satisfies the constraints (22.1.2). In the case of problems VLS1b and VLS2b the interior of the feasible region is quite small because the inequality constraints are not satisfied for a wide range of values of the variables. In such cases it may be almost as difficult to solve (22.3.1) as to solve the original problem.

The feasible starting points used for the test problems in this section are as follows:

For VLS1b: $x_1 = 3.22$, $x_2 = 4.52$
for VLS2b: $x_1 = 0.925$, $x_2 = -0.4712$
for PM1a(n): $x_i = 0.6$, $i = 1, \ldots, n$
for MMX1 and MMX2: $x_1 = x_2 = 0$, $x_3 = 1.1$
for OC4(n) and OC5(n) (n assumed to be even):
$x_1 \ldots x_\nu = 0.66$; $x_{\nu+1} \ldots x_n = -0.66$; $x_{n+1} = 1.5$ where $\nu = n/2$.
for MT1: $x_1 = 5$, $x_2 = 0.9$, $x_3 = 1.9$.

The progress made by B-SUMT on problem VLS1b is shown in Table 22.1. For comparison, this table also summarises the behaviour of P-SUMT from the same starting point. In both cases the unconstrained

	B-SUMT			P-SUMT		
k	$F(x_k)$	r_k	QNp Cost	$F(x_k)$	$\|c(x_k)\|$	QNp Cost
1	2.33×10^{-1}	1.0×10^{-1}	2/13	2.03×10^{-1}	1.8×10^{-2}	4/15
2	2.29×10^{-1}	2.5×10^{-2}	5/30	2.08×10^{-1}	7.1×10^{-3}	6/25
3	2.21×10^{-1}	6.3×10^{-3}	9/57	2.11×10^{-1}	2.1×10^{-3}	8/36
4	2.15×10^{-1}	1.6×10^{-3}	13/88	2.12×10^{-1}	5.4×10^{-4}	10/50
5	2.13×10^{-1}	3.9×10^{-4}	17/119	2.12×10^{-1}	1.4×10^{-4}	12/65
6	2.13×10^{-1}	9.8×10^{-5}	19/147	2.12×10^{-1}	3.4×10^{-5}	14/82
7	2.13×10^{-1}	2.4×10^{-5}	21/175	2.12×10^{-1}	8.6×10^{-6}	16/101
8	2.13×10^{-1}	6.1×10^{-6}	23/204			

Table 22.1.   B-SUMT and P-SUMT solutions to Problem VLS1b.

minimizer is QNp and the initial penalty parameter and rate of reduction are $r_0 = 0.1$, $\beta = 0.25$.

The penalty and barrier methods both use a similar number of unconstrained minimizations to solve VLS1b. However B-SUMT requires around 30% more quasi-Newton iterations and about twice as many function evaluations. This suggests that the log-barrier function is harder to minimize than the classical penalty function. In particular, doing a perfect line search in terms of $B(x, r)$ seems much more difficult than it is for $P(x, r)$. This is, at least in part, due to the requirement that $B(x, r)$ can only accept feasible points and so some trial steps during the line search have to be rejected.

If we use a weak line search by performing the unconstrained minimizations with QNw then B-SUMT converges in 39 iterations and 115 function calls and P-SUMT takes 24 iterations and 96 function calls.

Table 22.2 gives a broader comparison between the SUMT and SQP methods on a range of inequality constrained problems. Because B-SUMT appears in this table, the counts of iterations and function values are based on a feasible starting guess for each problem and so some of the entries for P-SUMT, AL-SUMT and AL-SQP differ from those in Table 21.2.

Method	VLS1b	VLS2b	PM1a(15)	MMX1	MMX2
B-SUMT/QNw	39/115	43/119	154/402	80/225	76/229
P-SUMT/QNw	24/96	21/57	95/283	54/222	52/225
AL-SUMT/QNw	16/64	10/20	56/139	30/106	34/129
AL-SQP	8/8	4/4	13/15	6/11	7/12

Method	OC4(6)	OC5(6)	MT1
B-SUMT/QNw	203/289	236/409	97/178
P-SUMT/QNw	59/160	51/177	47/161
AL-SUMT/QNw	50/128	45/145	32/94
AL-SQP	17/26	20/27	10/16

Table 22.2.   Performance of SUMT and SQP on Problems VLS1b–MT1.

We can see that the barrier function approach is usually the least competitive of the SUMT methods. Hence, in the form described in this chapter, its practical usefulness is normally confined to those problems where the function cannot be calculated at some infeasible points. A simple example would be if the expression for $F(x)$ included terms involving $\sqrt{x_i}$ because these are noncomputable if a constraint such as $x_i \geq 0$ is violated. In such situations it is important to use a method whose iterates stay inside the constraint boundaries.

In spite of the relatively poor performance of B-SUMT, the ideas behind the method are important because they are the foundation for the *interior point* methods described in the next chapter.

Exercises

1. Repeat the calculations in Table 22.2 using both low- and high-accuracy convergence tests and comment on the results, particularly in relation to the evidence of linear or superlinear convergence of the methods.

2. Use B-SUMT and other methods from Table 22.2 to solve a modified version of Problem TD1b in which $x_{max} = 3.5$ and $x_{min} = 1.8$.

3. Use B-SUMT with both a perfect and a weak line search to solve a version of Problem VLS1b in which $r_{max} = 0.28$.

4. Implement a version of B-SUMT which uses the reciprocal barrier function and investigate its performance on Problems TD1b and VLS2b.

5. Consider the Lagrange multiplier estimates provided by B-SUMT, P-SUMT and AL-SUMT at the solutions to MT1 and comment on any differences you observe.

6. Experiment with different choices of initial barrier parameter r_1 and scaling factor β in order to obtain the best performance of B-SUMT on Problems MT1 and PM1a(15).

7. Using any unconstrained minimization method, form and solve (22.3.1) to obtain feasible points for Problems TD2b, VLS2b and MT1.

8. Implement a spreadsheet version of the barrier SUMT method which uses SOLVER as the unconstrained minimizer and test it on Problems TD1b and VLS1b.

The British aircraft industry circa 1966 (Part 2) [4]

Oswald's a draughtsman with red hair and glasses
and a check shirt and beard and he's gone a bit soft
on the charms and the shape of the blonde buxom tracer
who lays out the spars and the ribs in the loft.

Recently made up to manager, Reginald
wears a black homburg, but you'd never guess
this big honey-bear man in crumpled blue trousers
is head of the office that calculates stress.

Chapter 23

Interior Point Methods

23.1.   Forming the transformed problem B-NLP

Interior point methods are related to barrier functions. They are widely used for nonlinear programming, following their introduction and continuing popularity as techniques for linear programming [40]. Consider the problem

$$\text{Minimize } F(x) \quad \text{subject to } c_i(x) \geq 0, \quad i = 1, \ldots, m. \qquad (23.1.1)$$

We can introduce additional slack variables to reformulate the inequalities as equalities and hence obtain a solution to (23.1.1) by finding x and w to solve

$$\text{Minimize } F(x) \qquad (23.1.2)$$

$$\text{s.t.} \quad c_i(x) - w_i = 0, \quad i = 1, \ldots, m \quad \text{and} \quad w_i \geq 0, \quad i = 1, \ldots, m.$$
$$(23.1.3)$$

If we deal with bounds on the w_i by a barrier term we obtain *Problem B-NLP* which involves a positive parameter, r.

$$\text{Minimize } F(x) - r \sum_{i=1}^{m} \log(w_i) \qquad (23.1.4)$$

$$\text{subject to } c_i(x) - w_i = 0, \quad i = 1, \ldots, m. \qquad (23.1.5)$$

The following result depends on fairly mild assumptions about F and the c_i.

Proposition Suppose $\{x^*, w^*, \lambda^*\}$ solves (23.1.3). If, for all r less than a constant ρ, Problem B-NLP has a unique solution $\hat{x}(r)$, $\hat{w}(r)$

M. Bartholomew-Biggs, *Nonlinear Optimization with Engineering Applications*,
DOI: 10.1007/978-0-387-78723-7_23, © Springer Science+Business Media, LLC 2008

with Lagrange multipliers $\hat{\lambda}(r)$ then

$$\{\hat{x}(r),\ \hat{w}(r),\ \hat{\lambda}(r)\} \rightarrow \{x^*,\ w^*,\ \lambda^*\}\ \text{as}\ r \rightarrow 0.$$

A sequential constrained minimization technique could be devised which solves B-NLP for a decreasing sequence of r-value in order to approach the solution of (23.1.3). However – as in AL-SQP – we would like to avoid the cost of complete minimizations by simply approximating solutions of B-NLP in a way that causes them to become more accurate as r approaches zero.

Exercises
1. Show that if $(x^*,\ w^*)$ is a solution of (23.1.2), (23.1.3) then x^* is also a solution of (23.1.1).
2. Form the problem B-NLP corresponding to problem VLS2b and solve it (e.g., by using SOLVER) for a decreasing sequence of values of the parameter r.
3. Derive an extension of problem B-NLP to deal with nonlinear programming problems that include both equality and inequality constraints.

23.2. Approximate solutions of Problem B-NLP

The Lagrangian function associated with B-NLP is

$$L(x, w, \lambda) = F(x) - r \sum_{i=1}^{m} \log(w_i) - \sum_{i=1}^{m} \lambda_i (c_i(x) - w_i). \qquad (23.2.1)$$

The first-order optimality conditions at the solution $(\hat{x},\ \hat{w},\ \hat{\lambda})$ are:

$$c_i(\hat{x}) - \hat{w}_i = 0, \quad i = 1, \dots, m; \qquad (23.2.2)$$

$$\nabla_x L = \nabla F(\hat{x}) - \sum_{i=1}^{m} \hat{\lambda}_i\, \nabla c_i(\hat{x}) = 0; \qquad (23.2.3)$$

$$\frac{\partial L}{\partial w_i} = -\frac{r}{\hat{w}_i} + \hat{\lambda}_i = 0, \quad i = 1, \dots, m. \qquad (23.2.4)$$

Equation (23.2.2) ensures feasibility. Equations (23.2.3) and (23.2.4) are stationarity conditions for the original variables and the slacks. In what follows, ∇ and ∇^2 operators without subscripts always relate to differentiation with respect to the original x variables only.

Suppose $(x,\ w,\ \lambda)$ is an approximate solution of B-NLP and we want to find $\delta x,\ \delta w,\ \delta\lambda$ so that $\hat{x} = x + \delta x$, $\hat{w} = w + \delta w$ and $\hat{\lambda} = \lambda + \delta\lambda$.

From (23.2.2)

$$c_i(x + \delta x) - w_i - \delta w_i = 0, \quad i = 1, \ldots, m$$

and a first-order Taylor approximation to $c_i(x + \delta x)$ gives

$$\nabla c_i(x)\delta x - \delta w_i = w_i - c_i(x), \quad i = 1, \ldots, m. \qquad (23.2.5)$$

From (23.2.3),

$$\nabla F(x + \delta x) - \sum_{i=1}^{m} (\lambda_i + \delta\lambda_i)\nabla c_i(x + \delta x) = 0$$

and by using first-order Taylor approximations of the gradient terms we get

$$\nabla F(x) + \nabla^2 F(x)\delta x - \sum_{i=1}^{m} (\lambda_i + \delta\lambda_i)(\nabla c_i(x) + \nabla^2 c_i(x)\delta x) = 0. \quad (23.2.6)$$

If we combine the terms in (23.2.6) which involve $\nabla^2 F$ and $\nabla^2 c_i$ and then ignore the second-order terms which feature the product $\delta\lambda_i\, \delta x$ we get

$$\tilde{G} = \nabla^2 F(x) - \sum_{i=1}^{m} \lambda_i \nabla^2 c_i(x).$$

From (23.2.6) we then obtain

$$\nabla F(x) + \tilde{G}\delta x - \sum_{i=1}^{m} (\lambda_i + \delta\lambda_i)\nabla c_i(x) = 0$$

which rearranges as

$$\tilde{G}\delta x - \sum_{i=1}^{m} \delta\lambda_i \,\nabla c_i(x) = \sum_{i=1}^{m} \lambda_i\, \nabla c_i(x) - \nabla F(x). \qquad (23.2.7)$$

Finally, from (23.2.4),

$$(w_i + \delta w_i)(\lambda_i + \delta\lambda_i) = r, \quad i = 1, \ldots, m.$$

Dropping the second-order term $\delta w_i \delta\lambda_i$ and rearranging we obtain

$$\delta w_i = \frac{r}{\lambda_i} - w_i - w_i \frac{\delta\lambda_i}{\lambda_i}, \quad i = 1, \ldots, m. \qquad (23.2.8)$$

Substituting for δw_i in (23.2.5) yields

$$\nabla c_i(x)\delta x + w_i \frac{\delta\lambda_i}{\lambda_i} = -c_i(x) + \frac{r}{\lambda_i}, \quad i = 1, \ldots, m. \qquad (23.2.9)$$

We now write $g = \nabla F(x)$ and let A denote the Jacobian matrix whose rows are $\nabla c_i(x)$, $i = 1, \ldots, m$. As usual, e denotes the m-vector with elements $e_i = 1$ and we let W, Λ be diagonal matrices whose elements are w_i and λ_i respectively. Then we can express (23.2.7) and (23.2.9) as a symmetric system of equations for δx and $\delta \lambda$. (These equations are somewhat similar to (19.2.9) and (19.2.10) which give δx and $\delta \lambda$ in augmented Lagrangian SQP.)

$$\tilde{G}\delta x - A^T \delta \lambda = -g + A^T \lambda \qquad (23.2.10)$$

$$-A\delta x - W\Lambda^{-1}\delta \lambda = c - r\Lambda^{-1}e. \qquad (23.2.11)$$

Once δx and $\delta \lambda$ have been found by solving (23.2.10), (23.2.11) we can recover δw from a rearrangement of (23.2.8)

$$\delta w = r\Lambda^{-1}e - w - W\Lambda^{-1}\delta \lambda. \qquad (23.2.12)$$

Later in this chapter we describe an algorithm based on (23.2.10)–(23.2.12) in which $(\delta x,\ \delta w)$ is regarded as a search direction along which an acceptable step must be determined.

An example

We consider the problem

$$\text{Minimize} \quad x_1^2 + 3x_2^2$$

subject to

$$x_1 + 5x - 2 - 1 \geq 0, \quad 5x_1 - x_2 - 0.25 \geq 0.$$

We start an iteration from $x = (0.25,\ 0.2)^T$ where $g = (0.5,\ 1.2)^T$ and $c = (0.25, 0.8)^T$. The Hessian and Jacobian matrices are

$$\tilde{G} = \begin{pmatrix} 2 & 0 \\ 0 & 6 \end{pmatrix} \quad \text{and} \quad A = \begin{pmatrix} 1 & 5 \\ 5 & -1 \end{pmatrix}.$$

We take $\lambda = (0.23, 0.054)^T$ as a starting guess because this gives $g \approx A^T\lambda$. A suitable choice for w can be based on the observation that, at a solution of B-NLP, $r = \lambda_i w_i$, for $i = 1, \ldots, m$. Therefore, if we set $r = 0.005$ we can take

$$w_i = \frac{0.005}{\lambda_i}, \quad \text{giving} \ w = (0.0217,\ 0.0926)^T.$$

Because

$$W = \begin{pmatrix} 0.0217 & 0 \\ 0 & 0.0926 \end{pmatrix} \quad \text{and} \quad \Lambda = \begin{pmatrix} 0.23 & 0 \\ 0 & 0.054 \end{pmatrix}$$

the equations (23.2.10), (23.2.11) for δx and $\delta \lambda$ are

$$\begin{pmatrix} 2 & 0 & -1 & -5 \\ 0 & 6 & -5 & 1 \\ -1 & -5 & -0.0945 & 0 \\ -5 & 1 & 0 & -1.7147 \end{pmatrix} \begin{pmatrix} \delta x_1 \\ \delta x_2 \\ \delta \lambda_1 \\ \delta \lambda_2 \end{pmatrix} = \begin{pmatrix} 0 \\ -0.1040 \\ 0.2283 \\ 0.7074 \end{pmatrix}.$$

Solving this system gives

$$\delta x_1 \approx -0.129, \quad \delta x_2 \approx -0.0197, \quad \delta \lambda_1 \approx -0.0126, \quad \delta \lambda_2 \approx -0.0489.$$

Hence the new point is $x \approx (0.1214, 0.1803)^T$ and the revised multipliers are $\lambda \approx (0.2174, 0.0051)^T$. Now from (23.2.12) we get

$$\delta w = 0.005 \begin{pmatrix} 4.348 \\ 18.52 \end{pmatrix} - \begin{pmatrix} 0.0217 \\ 0.0926 \end{pmatrix} - \begin{pmatrix} 0.0943 & 0 \\ 0 & 1.715 \end{pmatrix} \begin{pmatrix} -0.0126 \\ -0.0489 \end{pmatrix}.$$

This simplifies to $\delta w_1 \approx 0.0012$, $\delta w_2 \approx 0.0839$ and so the corrected slack variables are $w \approx (0.0229, 0.1765)^T$.

The solution of the original problem is at $x^* \approx (0.1071, 0.1786)^T$ with the first constraint binding but not the second. Hence the iteration has moved the variables appreciably closer to x^* and has also moved λ_2 closer to zero.

Exercises

1. Use (23.2.8) to eliminate $\delta \lambda$ instead of δw and obtain equations similar to (23.2.10), (23.2.11) with δx and δw as unknowns. By performing a suitable change of variable show that this can be made into a symmetric system.

2. If the inverse $\tilde{G}^{-1}$ is available, show that values of δx and $\delta \lambda$ which solve (23.2.10), (23.2.11) can be obtained from

$$\delta \lambda = (A\tilde{G}^{-1}A^T + W\Lambda^{-1})^{-1}(A\tilde{G}^{-1}g + r\Lambda^{-1}e - c - A\tilde{G}A^T\lambda)$$

$$\delta x = \tilde{G}^{-1}(-g + A^T(\lambda + \delta \lambda)).$$

3. Do a second iteration of the worked example given above. What would have happened on the first iteration if we had chosen $r = 0.0025$?

4. Form and solve equations (23.2.10), (23.2.11) to obtain δx, δw and $\delta \lambda$ for the problem

$$\text{Minimize} \quad x_1^2 + 2x_2^2 \quad \text{s.t.} \quad x_1 + x_2 \geq 1$$

starting from the values $x_1 = x_2 = 1$, $w_1 = 0.1$, $\lambda_1 = 1$, $r = 0.1$.

23.3.   An interior point algorithm

We can now give an outline of an interior point algorithm. The new values x, w obtained on iteration k will be of the form

$$x_{k+1} = x_k + s\delta x_k, \quad w_{k+1} = w_k + s\delta w_k$$

where δx_k and δw_k are obtained by solving (23.2.10), (23.2.11) with all the coefficients and right-hand side values evaluated at (x_k, w_k, λ_k). The steplength s must be small enough for $w_k + s\delta w_k$ to be positive because we are only interested in feasible points of subproblem B-NLP. We can find an upper limit $\bar{s}$ on the stepsize from a formula such as

$$\bar{s} = \min\left(-0.9\frac{w_{k_i}}{\delta w_{k_i}}\right) \tag{23.3.1}$$

where the minimum is taken over all i such that $\delta w_{k_i} < 0$. (There is no upper limit on stepsize if all the elements of δw_k are nonnegative.)

We also need to choose s (less than $\bar{s}$) so that $(x_k + s\delta x_k, w_k + s\delta w_k)$ is a better solution estimate than (x_k, w_k). We could, for example, perform a line search in terms of an augmented Lagrangian for problem B-NLP, namely

$$\hat{M}(x, w, v, r) = F(x) - r\sum_{i=1}^{m}\log(w_i) - (c(x) - w)^T v + \frac{||c(x) - w||_2^2}{r} \tag{23.3.2}$$

where v is a vector of Lagrange multiplier approximations. If $\delta\lambda_k$ is obtained by solving (23.2.10), (23.2.11) then it can be shown (under certain circumstances) that the choice $v = \lambda_k + \delta\lambda_k$ ensures $(\delta x_k, \delta w_k)$ is a descent direction for (23.3.2) at (x_k, w_k). This can be expressed as

$$\delta x_k^T \nabla_x \hat{M}(x_k, w_k, v, r) + \delta w_k^T \nabla_w \hat{M}(x_k, w_k, v, r) < 0. \tag{23.3.3}$$

However, when x_k is far from x^*, it is not clear that

$$v = \lambda_{k+1} = \lambda_k + \delta\lambda_k$$

is a sufficiently good Lagrange multiplier estimate to ensure that a line search with respect to (23.3.2) will be helpful for overall convergence. In fact we must restrict Lagrange multiplier estimates to strictly positive values, since Λ^{-1} will not exist if any of the λ_i is zero. Therefore instead of using the full correction $\delta\lambda_{k_i}$ to update the ith multiplier we employ a modified formula

$$\lambda_i^+ = \max(\lambda_{\min}, \lambda_{k_i} + \delta\lambda_{k_i}) \tag{23.3.4}$$

where $\lambda_{\min}$ is a small positive threshold value.

The algorithm IPM, given below, resembles AL-SQP in that it fixes "sensible" values for the v appearing in (23.3.2) and retains them until tests on the errors in optimality conditions indicate that they can be replaced. When $v \neq \lambda_k + \delta\lambda_k$ the descent property (23.3.3) may be ensured if r is chosen sufficiently small on each iteration.

Algorithm IPM uses a quasi-Newton approach to update a matrix B_k to approximate $\tilde{G}$ (which is an estimate of the Hessian of the Lagrangian function). Revision of λ and r is done as in AL-SQP using an error function based on the optimality conditions (23.2.2)–(23.2.4), namely

$$\tau(x, w, \lambda, r) = ||c(x) - w||^2 + ||g - A^T\lambda||^2 + ||W\Lambda e - re||^2. \quad (23.3.5)$$

When τ is sufficiently small we can assume we are close enough to a solution of Problem B-NLP to permit the penalty parameter r to be reduced.

Interior point algorithm (IPM)

Choose initial values x_0, $w_0 (> 0)$, $\lambda_0 (> 0)$, B_0, r_0 and v_0
Choose a scaling factor $\beta < 1$ and set $\tau_r^- = \tau(x_0, w_0, \lambda_0, r_0)$
Repeat for $k = 0, 1, 2, \ldots$
Obtain δx_k and $\delta\lambda_k$ by solving

$$B_k\delta x - A_k^T\delta\lambda = -g_k + A_k^T\lambda_k$$

$$-A_k\delta x - W_k\Lambda_k^{-1}\delta\lambda = c_k - r_k\Lambda_k^{-1}e$$

Set $\delta w_k = r_k\Lambda_k^{-1}e - w_k - W_k\Lambda_k^{-1}\delta\lambda_k$, and $\bar{s} = 1$
Find $\bar{s}$ using (23.3.1) and λ^+ using (23.3.4).
Get $x_{k+1} = x_k + s\delta x_k$, $w_{k+1} = w_k + s\delta w_k$ $(s \leq \bar{s})$, by a line search to give

$$\hat{M}(x_{k+1}, w_{k+1}, v_k, r_k) < \hat{M}(x_k, w_k, v_k, r_k)$$

Obtain B_{k+1} by a quasi-Newton update of B_k
Set $\lambda_{k+1} = \lambda_k$, $r_{k+1} = r_k$ and $v_{k+1} = v_k$
If $\tau(x_{k+1}, w_{k+1}, \lambda^+, r_k) < \tau_r^-$ then
set $\tau_r^- = \tau(x_{k+1}, w_{k+1}, \lambda^+, r_k)$, $\lambda_{k+1} = \lambda^+$, $r_{k+1} = \beta r_k$, $v_{k+1} = \lambda_{k+1}$
until $||\tau(x_{k+1}, w_{k+1}, \lambda_{k+1}, 0)||$ is sufficiently small.

Each iteration of IPM is based on estimating a solution of Problem B-NLP. Many variations of this one central idea have been suggested, leading to algorithms which get δx, δw and $\delta\lambda$ from equations which are somewhat different from (23.2.10)–(23.2.12) and which perform line-searches using merit functions other than (23.3.2). Discussion of such alternative algorithms can be found in [26] and [67], for instance.

Exercises

1. Suppose δx, δw, $\delta \lambda$ are obtained by solving (23.2.10)–(23.2.12). In order for $(\delta x, \delta w)$ to be a descent direction for the augmented Lagrangian $\hat{M}$ given by (23.3.2) we require

$$\delta x^T \nabla_x \hat{M} + \delta w^T \nabla_w \hat{M} < 0$$

$$\text{where} \quad \nabla_x \hat{M} = g + \frac{2}{r} A^T \{ 2(c - w) - v \}$$

$$\text{and} \quad \nabla_w \hat{M} = -r W^{-1} e + v - \frac{2}{r}(c - w).$$

Show that $\delta x^T \nabla_x \hat{M} + \delta w^T \nabla_w \hat{M}$ is equivalent to the expression

$$-\delta x^T \tilde{G} \delta x - \delta w^T \Lambda W^{-1} \delta w - \frac{2}{r}(c - w)^T (c - w) - (c - w)^T (\lambda^+ - v)$$

where $\lambda^+ = \lambda + \delta \lambda$.

2. Use the result of the previous exercise to show that, if $\tilde{G}$ is positive-definite and if $W^{-1}\Lambda$ is positive semi-definite then $(\delta x, \ \delta w)$ is a descent direction with respect to (23.3.2) for any value of r if $v = \lambda^+$. Show also that if $v \neq \lambda^+$ the descent property with respect to $\hat{M}$ may be ensured if r is chosen sufficiently small on each iteration.

3. Explain why (23.2.4) implies that, when the parameter r_k is replaced by βr_k, a good way to adjust the values of the slack variables w and the multipliers λ might be to use one of the following formulae for each $i = 1, \ldots, m$.

$$\text{if } \lambda_{k_i} < w_{k_i} \quad \text{then } \lambda_{(k+1)_i} = \beta \lambda_{k_i} \quad \text{and} \quad w_{(k+1)_i} = w_{k_i}$$

$$\text{else} \quad w_{(k+1)_i} = \beta w_{k_i} \quad \text{and} \quad \lambda_{(k+1)_i} = \lambda_{k_i}.$$

23.4. Results with IPM

The OPTIMA implementation of the interior point approach is called IPM and is based on a variation [7] of the algorithm given in the previous section. Table 23.1 shows progress of IPM iterations on Problem VLS1b when $r_1 = 0.1$ and $\beta = 0.25$. Comparison with Table 22.1 shows that IPM converges more quickly than both B-SUMT and P-SUMT. This happens because IPM avoids explicit minimizations of the subproblems for each value of barrier parameter r.

Table 23.2 compares the number of iterations and function calls needed by IPM and B-SUMT on a number of test problems. The unconstrained minimizations in B-SUMT use a weak line search. Clearly IPM is much

k	$F(x_k)$	r_k	itns/function calls
1	3.71×10^{-1}	1.0×10^{-1}	1/2
2	2.59×10^{-1}	2.5×10^{-2}	3/4
3	2.18×10^{-1}	6.25×10^{-6}	6/8
4	2.15×10^{-1}	1.56×10^{-3}	10/13
5	2.13×10^{-1}	3.91×10^{-4}	14/17
6	2.13×10^{-1}	9.5×10^{-5}	18/21

Table 23.1. IPM solution to Problem VLS1b.

Method	VLS1b	VLS2b	PM1a(15)	MMX1	MMX2
IPM	18/21	27/37	15/18	19/21	20/24
B-SUMT	39/115	43/119	154/402	80/225	76/229

Method	OC4(6)	OC5(6)	MT1
IPM	29/30	58/80	25/26
B-SUMT	203/289	236/409	97/178

Table 23.2. Performance of IPM and B-SUMT on Problems VLS1b–MT1.

more efficient than B-SUMT in all cases and we can conclude that the basic idea of proceeding via approximate solutions of Problem B-NLP is better than performing a sequence of accurate minimizations of a barrier function $B(x, r)$.

The entries in Table 23.2 show that the OPTIMA implementation of IPM is sometimes – but by no means always – competitive with AL-SQP and SOLVER. It is important not to draw too sweeping a conclusion from this regarding the general merits of interior point and SQP methods. Since the 1980s, interior point methods have been the subject of much research and many algorithms have been proposed. Some are designed for special situations such as LP or QP problems. Those intended for the general (possibly nonconvex) nonlinear programming problem include [13] and [28]. Some of these implementations of interior point methods are more sophisticated than IPM and include features for accelerating convergence which make them much more competitive with the SQP approach. Hence it still seems an open question which of these two techniques is "better".

The computational cost of an IPM iteration can be similar to that for AL-SQP, because both methods get a search direction by solving a linear system obtained by approximating the optimality conditions for a perturbed form of the original minimization problem. The IPM system will include all the inequality constraints and so will usually be larger than the system used by AL-SQP which only involves constraints in the current active set. On the other hand, AL-SQP may have to solve several systems on each iteration until the correct active set is established. It is suggested in [31] that IPM and SQP can co-exist because IP algorithms can be an efficient way to solve the QP subproblems in SQP methods.

Exercises

1. Investigate the sensitivity of IPM to changes in the choices of r_1 and β.

2. By choosing starting guesses which differ by only 1%, 5%, ... from the exact solutions of the problems in Table 23.2, determine how competitive IPM can be with AL-SQP in the neighbourhood of x^*.

3. Apply IPM to a variant of Problem TD1b in which the constraint on volume is expressed as $x_1 x_2 x_3 \geq V^*$.

4. Apply IPM to a variant of Problem MT1 in which the cutting depth is fixed as $d = 1$.

5. Extend the comparison in Table 23.2 to include OC4(n) and OC5(n) for values of $n > 6$ and compare the execution times of IPM and B-SUMT. Does B-SUMT become more or less competitive if perfect line searches are used?

Chapter 24

A Summary of Constrained Methods

To summarise the work covered in the preceding chapters we give a checklist of the properties of the constrained optimization methods that have been described.

Quadratic programming
Only used for quadratic F with linear c_i.
Simply solves the KKT equations when constraints are equalities.
Uses active set approach for inequality constraints, which means it solves a sequence of equality constrained problems.
Approaches solution via a sequence of feasible points.

Reduced-gradients
Works best for linear constraints.
Uses constraints to eliminate t variables and then does an unconstrained step in the other $n - t$ variables.
Can also work for nonlinear constraints but then needs restoration steps to regain feasibility.
Implemented in Excel SOLVER.
Approaches solution via a sequence of feasible points.

Penalty function SUMT
$P(x, r)$ adds squared constraint violations to F.
P is minimized for a decreasing sequence of r-values.
Minima of P converge to constrained solution and approach solution via sequence of infeasible points.
Can be better than reduced-gradients for nonlinear constraints.
Can have numerical difficulties as $r \to 0$.

M. Bartholomew-Biggs, *Nonlinear Optimization with Engineering Applications*,
DOI: 10.1007/978-0-387-78723-7_24, © Springer Science+Business Media, LLC 2008

Augmented Lagrangian SUMT

$M(x, v, r)$ is formed from $P(x, r)$ by including extra linear term involving violated constraints.

M is minimized for a sequence of values of parameters r and v.

Minima of $M \to$ constrained solution if the $v \to$ Lagrange multipliers

No need for $r \to 0$ so M does not have the same numerical difficulties as P.

Approaches solution via a sequence of infeasible points.

Usually more efficient than penalty function SUMT.

Sequential quadratic programming

Makes a QP on every iteration with a quadratic model of F and linearised c_i.

Solves this QP subproblem to get a search direction.

Chooses a new point by a weak line search in terms of the augmented Lagrangian or other penalty function.

Approaches solution via a sequence of infeasible points.

More efficient than reduced-gradients or SUMT when constraints are non-linear.

Competitive with reduced gradients for linear c_i.

Barrier function SUMT

Works for problems with inequality constraints only.

$B(x, r)$ includes barrier term involving reciprocals (or logs) of constraints.

B is minimized for a decreasing sequence of r-values.

Minima of B tend to constrained solution.

Approaches solution via a sequence of feasible points.

Usually less efficient than other SUMT methods but is still useful if F is not computable at infeasible points.

Interior point method

Uses slack variables to turn inequalities to equalities.

Handles slack-variable positivity by a barrier term.

Avoids cost of SUMT by only approximating minima of barrier function.

Can be competitive with SQP for nonlinear constraints.

Competitive with reduced-gradients for linear c_i.

Is an alternative to active-set approach for inequality constrained QP.

Chapter 25

The OPTIMA Software

25.1.   Accessing OPTIMA

The OPTIMA fortran90 codes can be obtained from the Web via the
ftp site `ftp.feis.herts.ac.uk/pub/matqmb/OPTIMA`. The codes can also be
obtained by anonymous ftp using a UNIX dialogue as in the following
example for getting the problem TD1 (user inputs are underlined).

```
ftp ftp.feis.herts.ac.uk

Connected to ftp.feis.herts.ac.uk.
Welcome to EIS at the University of Hertfordshire.

Name (ftp.feis.herts.ac.uk:comqmb): ftp
331 Please specify the password.
Password: ftp

... University of Hertfordshire logo and welcome appears here ...
230 Login successful.

ftp> cd pub/matqmb/OPTIMA
250 Directory successfully changed.
ftp> ls
... a list of the available codes appears here ...
ftp> get TD1.f90

local: TD1.f90 remote: TD1.f90
200 PORT command successful.
150 Opening BINARY mode data connection for VLS2.f90 (1505 bytes).
226 File send OK.
1505 bytes received in 0.00 secs (1595.8 kB/s)
```

M. Bartholomew-Biggs, *Nonlinear Optimization with Engineering Applications*,
DOI: 10.1007/978-0-387-78723-7_25, © Springer Science+Business Media, LLC 2008

25.2.   Running OPTIMA

The OPTIMA software has been developed on a Sun workstation and it also runs on a PC under Visual Fortran. The code is intended to be portable but minor changes may be needed before it will compile and run on other systems.

In order to run, for example, the tank design problem TD1 it is necessary to compile and link the program file TD1.f90 and the file OPTIMA.f90 which includes modules MINPAC and OPFAD. These contain, respectively, the procedures for optimization and automatic differentiation. On running the resulting executable file, the user will be able to make choices about the solution technique as illustrated by the following dialogue. (User inputs appear slightly to the left of program output text.)

```
Problem TD1
Choose optimization method:
univariate search (1); DIRECT(2)
Steepest descent(3); Newton(4); quasi-Newton(5); conjugate gradients(6)
5
Use weak line search (y/n)?
y
 Solution accuracy? Low(L); Standard(S); High(H)
s
 Quasi-Newton (weak search and    mid tolerance)
 Converged after      9 iterations and    13 function calls
 Solution x =
  0.170997E+01   0.341997E+01
 with function value F =    0.350882E+02
```

25.3.   Modifying and creating test problems

It is expected that many users will simply compile and run the example programs using the built-in choices illustrated in the previous section. Useful experience can be obtained by treating the given codes as "black boxes" for demonstrating the behaviour of different methods. (Note that the quoted figures for numbers of iterations and function calls may not be exactly replicated when a user runs a particular example because OPTIMA software may undergo periodic revisions.)

Some readers, however, may wish to make small modifications such as changing a starting guess or altering some of the parameters in a problem. Such minor changes can be probably be made by those with no previous Fortran experience. A few, more ambitious, users may wish to formulate and solve their own problems. In order to facilitate both possibilities we give below some program listings to serve as templates for those who wish to pose their own example problems.

An unconstrained problem

The first listing is for the solution of Problem TD1. The routine OPTIMIZE1 provides the interface to the optimization routine which offers some user choices about the solution technique. Comments in the listing show where changes could be made to alter the starting guess or the target volume of the tank.

```fortran
PROGRAM TD1
! *** Main program for tank design problem ***
  USE minpac
  IMPLICIT NONE
  REAL*8, DIMENSION(:), ALLOCATABLE :: x
  INTEGER :: n, method
  INTERFACE
  SUBROUTINE calfun(n,x,f,g)
  INTEGER, INTENT(in) :: n
  REAL*8, INTENT(in), DIMENSION(1:n) :: x
  REAL*8, INTENT(out)  :: f
  REAL*8, INTENT(out), DIMENSION(1:n) :: g
  END SUBROUTINE calfun
  END INTERFACE
  PRINT"(' Problem TD1')"
  ! *** set number of variables and starting point ***
  n = 2;  ALLOCATE(x(1:n)); x= (/2.0D0, 2.0D0/)
  CALL suppress_minpac_history;  CALL set_minpac_iterations(5000);
  CALL OPTIMIZE1(n,x,method,calfun);  DEALLOCATE(x)
END PROGRAM TD1

SUBROUTINE TD1fg(x,f,g)
! *** user-supplied function and gradient for tank design problem ***
  USE opfad
  REAL*8, DIMENSION(1:2), INTENT(in) :: x
  REAL*8, INTENT(out) :: f
  REAL*8, DIMENSION(1:2), INTENT(out) :: g
  TYPE(doublet) :: xx(2), ff
  REAL*8 :: Vstar
  Vstar = 20.0D0   ! *** set target volume ***
  CALL INITIALIZE(2,x,xx)  ! *** convert real variables to doublet form ***
  ff = 2.0D0*xx(1)*xx(2) + 2.0D0*vstar/xx(2) + Vstar/xx(1)
  f = VALUE(ff); g = GRADIENT(ff)  ! *** extract function value and gradient ***
END SUBROUTINE TD1fg

! *** General-purpose interface routines ***
SUBROUTINE calfun(n,x,f,g)
  USE opfad;
  IMPLICIT NONE
  INTEGER, INTENT(in) :: n
  REAL*8, INTENT(in), DIMENSION(1:n) :: x
  REAL*8, INTENT(out)  :: f
  REAL*8, INTENT(out), DIMENSION(1:n) :: g
  CALL TD1fg(x,f,g)
END SUBROUTINE calfun

FUNCTION funval(n, x)
  IMPLICIT NONE
```

```
  INTEGER, INTENT(in) :: n
  REAL*8, INTENT(in), DIMENSION(1:n) :: x
  REAL*8 :: funval, f
  REAL*8, DIMENSION(1:2) :: g
  CALL TD1fg(x,f,g); funval = f
END FUNCTION funval
```

Much of the above listing could remain unaltered if the reader wanted to pose a different unconstrained minimization problem. A different expression for the objective function would have to appear in the body of the function TD1fg. There might also have to be changes to the value of n and the starting values for the variables.

An equality constrained problem

A second example shows a program to set up Problem TD1a which has only equality constraints. Note that the interface with OPTIMA is now through the subroutine OPTIMIZE3. As with the previous example, the comments in the listing show where changes to the problem might be made. To generate a new problem it would be necessary to put new expressions for the function and constraints in the body of TD1cfg. The dimension statements in the main program and in the subroutine Sumtfun need to be in agreement with the numbers of variables and constraints in the new problem

```
PROGRAM TD1a
  USE minpac
  IMPLICIT NONE
  REAL*8, DIMENSION(:), ALLOCATABLE :: x
  INTEGER :: n, method, m, me
  COMMON/sumt/method
  INTERFACE
  SUBROUTINE calfun(x,n,m,f,c,g,A)
  USE opfad;
  IMPLICIT NONE
  INTEGER, INTENT(in) :: n,m
  REAL*8, INTENT(in), DIMENSION(1:n) :: x
  REAL*8, INTENT(out)  :: f
  REAL*8, INTENT(out), DIMENSION(1:m) :: c
  REAL*8, INTENT(out), DIMENSION(1:n), OPTIONAL :: g
  REAL*8, INTENT(out), DIMENSION(1:m,1:n), OPTIONAL :: A
  END SUBROUTINE calfun
  SUBROUTINE Sumtfun(n,x,P,gradP)
  USE minpac
  INTEGER, INTENT(in) :: n
  REAL*8, DIMENSION(1:n), INTENT(in) :: x
  REAL*8, INTENT(out) :: P
  REAL*8, DIMENSION(1:n), INTENT(out) :: gradP
  END SUBROUTINE Sumtfun
  END INTERFACE
  PRINT"(' Problem TD1a')"
  ! *** Set number of variables and total number of constraints ***
```

```
   n = 3; m=1
! *** Set number of equality constraints ***
   me = 1
! *** Set starting guess ***
   ALLOCATE(x(1:n)); x= (/2.0D0, 2.0D0, 5.0D0/)
   CALL print_minpac_history;    CALL set_minpac_iterations(5000);
   CALL set_initial_penalty(0.1D0); CALL set_penalty_scaling(0.25D0)
   CALL OPTIMIZE3(n,x,m,me,method,Sumtfun,calfun);   DEALLOCATE(x)
END PROGRAM TD1a

SUBROUTINE TD1cfg(x,f,c,g,A)
! *** user-supplied function and constraints for tank design problem TD1a ***
   USE opfad
   REAL*8, DIMENSION(1:3), INTENT(in) :: x
   REAL*8, INTENT(out) :: f
   REAL*8, DIMENSION(1:1), INTENT(out) :: c
   REAL*8, DIMENSION(1:3), INTENT(out) :: g
   REAL*8, DIMENSION(1:1,1:3), INTENT(out) :: A
   TYPE(doublet) :: xx(3), ff, cc(1)
   REAL*8 :: Vstar
   Vstar = 20.0D0; ! *** target value for volume ***
   CALL INITIALIZE(3,x,xx)   ! *** convert real variables to doublet form ***
! *** evaluate function in doublet form ***
   ff = 2.0D0*xx(1)*xx(2) + 2.0D0*xx(1)*xx(3) + xx(2)*xx(3)
! *** evaluate constraint in doublet form ***
   cc(1) = xx(1)*xx(2)*xx(3) - Vstar
   f = VALUE(ff); g =GRADIENT(ff) ! *** extract function values and gradients ***
   c = VALUES(cc,1); A = NORMALS(cc,1) ! *** extract constraints and Jacobian ***
END SUBROUTINE TD1cfg

SUBROUTINE calfun(x,n,m,f,c,g,A)
   USE opfad;
   IMPLICIT NONE
   INTEGER, INTENT(in) :: n,m
   REAL*8, INTENT(in), DIMENSION(1:n) :: x
   REAL*8, INTENT(out)   :: f
   REAL*8, INTENT(out), DIMENSION(1:m) :: c
   REAL*8, INTENT(out), DIMENSION(1:n), OPTIONAL :: g
   REAL*8, INTENT(out), DIMENSION(1:m,1:n), OPTIONAL :: A
   REAL*8, DIMENSION(1:n) :: gdum
   REAL*8, DIMENSION(1:m,1:n) :: Adum
   CALL TD1cfg(x,f,c,gdum,Adum)
   IF(PRESENT(g))g=gdum; IF(PRESENT(A))A=Adum
END SUBROUTINE calfun

SUBROUTINE Sumtfun(n,x,P,gradP)
 USE minpac
 INTEGER, INTENT(in) :: n
 REAL*8, DIMENSION(1:n), INTENT(in) :: x
 REAL*8, INTENT(out) :: P
 REAL*8, DIMENSION(1:n), INTENT(out) :: gradP
 REAL*8 :: f
 REAL*8, DIMENSION(1:n) :: g
! *** NB dimension of c and first dimension of A must be number of
            constraints ***
 REAL*8, DIMENSION(1:1) :: c
 REAL*8, DIMENSION(1:1,1:n) :: A
```

```
INTEGER :: method
COMMON/sumt/method
CALL TD1cfg(x,f,c,g,A)
! *** Parameter 5 of Make_P and Make_AL must be number of equality
        constraints ***
IF (method == 1)CALL Make_P(f,g,c,A,1,P,gradP)
IF (method == 2)CALL Make_AL(f,g,c,A,1,P,gradP)
END SUBROUTINE Sumtfun
FUNCTION funval(n,y)
funval = 0.0
END FUNCTION funval
```

An equality and inequality constrained problem

The third code listing is for TD1b which is a problem with a mixture
of equality and inequality constraints. This also uses OPTIMIZE3 as the
interface to OPTIMA. In order to generate new problems of this type,
a user would need to write a new body for the subroutine TD1cfg and
ensure that correct values are assigned to the variables representing n,
m (the total number of constraints) and m_e (the number of equality
constraints). Note that the constraints must be numbered so that the
equalities are the first m_e elements in the vector $c_1, \ldots, c_m$. Note also
that the dimension statements in the subroutine Sumtfun must agree with
the actual number of constraints.

```
PROGRAM TD1b
! *** Main program for tank design problem with inequality constraints ***
  USE minpac
  IMPLICIT NONE
  REAL*8, DIMENSION(:), ALLOCATABLE :: x
  INTEGER :: n, method, m, me
  COMMON/sumt/method
  INTERFACE
  SUBROUTINE calfun(x,n,m,f,c,g,A)
  USE opfad;
  IMPLICIT NONE
  INTEGER, INTENT(in) :: n,m
  REAL*8, INTENT(in), DIMENSION(1:n) :: x
  REAL*8, INTENT(out)  :: f
  REAL*8, INTENT(out), DIMENSION(1:m) :: c
  REAL*8, INTENT(out), DIMENSION(1:n), OPTIONAL :: g
  REAL*8, INTENT(out), DIMENSION(1:m,1:n), OPTIONAL :: A
  END SUBROUTINE calfun
  SUBROUTINE Sumtfun(n,x,P,gradP)
  USE minpac
  INTEGER, INTENT(in) :: n
  REAL*8, DIMENSION(1:n), INTENT(in) :: x
  REAL*8, INTENT(out) :: P
  REAL*8, DIMENSION(1:n), INTENT(out) :: gradP
  END SUBROUTINE Sumtfun
  END INTERFACE
  PRINT"(' Problem TD1b')"
  ! *** Set number of variables and total number of constraints ***
```

```
    n = 3; m=4
! *** Set number of equality constraints ***
    me = 1
! *** Set starting guess ***
    ALLOCATE(x(1:n)); x= (/2.0D0, 2.0D0, 5.0D0/)
    CALL print_minpac_history;  CALL set_minpac_iterations(5000);
    CALL set_initial_penalty(0.1D0); CALL set_penalty_scaling(0.25D0)
    CALL OPTIMIZE3(n,x,m,me,method,Sumtfun,calfun);  DEALLOCATE(x)
END PROGRAM TD1b

SUBROUTINE TD1cfg(x,f,c,g,A)
! *** user-supplied function and constraints for tank design problem with
            bounds ***
    USE opfad
    REAL*8, DIMENSION(1:3), INTENT(in) :: x
    REAL*8, INTENT(out) :: f
    REAL*8, DIMENSION(1:4), INTENT(out) :: c
    REAL*8, DIMENSION(1:3), INTENT(out) :: g
    REAL*8, DIMENSION(1:4,1:3), INTENT(out) :: A
    TYPE(doublet) :: xx(3), ff, cc(4)
    REAL*8 :: Vstar
    Vstar = 20.0D0; ! *** target volume ***
    CALL INITIALIZE(3,x,xx) ! *** convert real variables to doublet form ***
! *** evaluate function in doublet form ***
    ff = 2.0D0*xx(1)*xx(2) + 2.0D0*xx(1)*xx(3) + xx(2)*xx(3)
! *** evaluate constraints in doublet form (equalities always first) ***
    cc(1) = xx(1)*xx(2)*xx(3) - Vstar
    do k = 1,3
      cc(k+1) = xx(k) - 1.9D0
    end do
    f = VALUE(ff);  g = GRADIENT(ff)! *** extract function and gradient ***
    c = VALUES(cc,4); A = NORMALS(cc,4) ! *** extract constraints and Jacobian ***
END SUBROUTINE TD1cfg

SUBROUTINE calfun(x,n,m,f,c,g,A)
  USE opfad;
  IMPLICIT NONE
  INTEGER, INTENT(in) :: n,m
  REAL*8, INTENT(in), DIMENSION(1:n) :: x
  REAL*8, INTENT(out)  :: f
  REAL*8, INTENT(out), DIMENSION(1:m) :: c
  REAL*8, INTENT(out), DIMENSION(1:n), OPTIONAL :: g
  REAL*8, INTENT(out), DIMENSION(1:m,1:n), OPTIONAL :: A
  REAL*8, DIMENSION(1:n) :: gdum
  REAL*8, DIMENSION(1:m,1:n) :: Adum
  CALL TD1cfg(x,f,c,gdum,Adum)
  IF(PRESENT(g))g=gdum; IF(PRESENT(A))A=Adum
END SUBROUTINE calfun

SUBROUTINE Sumtfun(n,x,P,gradP)
  USE minpac
  INTEGER, INTENT(in) :: n
  REAL*8, DIMENSION(1:n), INTENT(in) :: x
  REAL*8, INTENT(out) :: P
  REAL*8, DIMENSION(1:n), INTENT(out) :: gradP
  REAL*8 :: f
  REAL*8, DIMENSION(1:n) :: g
```

```
! *** dimension of c and first dimension of A must be total number of
          constraints ***
REAL*8, DIMENSION(1:4) :: c
REAL*8, DIMENSION(1:4,1:n) :: A
INTEGER :: method
COMMON/sumt/method
CALL TD1cfg(x,f,c,g,A)
! *** Parameter 5 of make_P and Make_AL must be number of equality
          constraints ***
IF (method == 1)CALL Make_P(f,g,c,A,1,P,gradP)
IF (method == 2)CALL Make_AL(f,g,c,A,1,P,gradP)
END SUBROUTINE Sumtfun
FUNCTION funval(n,y)
funval = 0.0
END FUNCTION funval
```

A sum-of-squares problem

The next listing shows how to set up the sum-of-squares unconstrained problem VLS2. This differs from the code for the general unconstrained example TD1 in that the user-supplied routine (VLS2fg) must calculate both the gradient of F and also the Jacobian matrix of the subfunctions. The interface to OPTIMA is via the routine OPTIMIZE2 which offers the Gauss–Newton method as an additional solver option. Coding of a new problem would require the replacement of data and expressions in subroutine VLS2fg.

```
PROGRAM VLS2
! *** Main program for data-fitting problem VLS2 ***
  USE minpac
  IMPLICIT NONE
  REAL*8, DIMENSION(:), ALLOCATABLE :: x
  INTEGER :: n, method, m
  INTERFACE
  SUBROUTINE calfun(n,x,f,g)
  INTEGER, INTENT(in) :: n
  REAL*8, INTENT(in), DIMENSION(1:n) :: x
  REAL*8, INTENT(out)  :: f
  REAL*8, INTENT(out), DIMENSION(1:n) :: g
  END SUBROUTINE calfun
  END INTERFACE
  PRINT"(' Problem VLS2')"
! *** Set number of variables and number of terms in sum-of-squares ***
  n = 2; m=5
! *** Set starting guess ***
  ALLOCATE(x(1:n)); x= (/0.0D0, 0.0D0/)
  CALL suppress_minpac_history;  CALL set_minpac_iterations(5000);
  CALL OPTIMIZE2(n,m,x,method,calfun); DEALLOCATE(x)
END PROGRAM VLS2

SUBROUTINE VLS2fg(x,f,g)
! user-supplied function gradient and Jacobian evaluation for problem VLS2
  USE opfad; USE minpac
  REAL*8, DIMENSION(1:2), INTENT(in) :: x
```

```
   REAL*8, INTENT(out) :: f
   REAL*8, DIMENSION(1:2), INTENT(out) :: g
   TYPE(doublet) :: xx(2), ff, ss(5)
   REAL*8 :: t(5),z(5)
   INTEGER :: k
   t = (/0.0,1.0,2.0,3.0,4.0/); z = (/1.0,0.5,0.4,0.3,0.2/)
! *** set data values ***
   CALL INITIALIZE(2,x,xx) ! *** convert real variables to doublet form ***
   ff = 0.0D0
! *** evaluate terms of sum of squares and accumulate their sum ***
   DO k = 1,5
      ss(k) = (xx(1)*exp(xx(2)*t(k)) - z(k)); ff = ff + ss(k)**2
   END DO
! *** extract function value and gradient and Jacobian of subfunctions ***
   f = VALUE(ff);   g = GRADIENT(ff); Jac = normals(ss,5)
END SUBROUTINE VLS2fg

SUBROUTINE calfun(n,x,f,g)
   USE opfad;
   IMPLICIT NONE
   INTEGER, INTENT(in) :: n
   REAL*8, INTENT(in), DIMENSION(1:n) :: x
   REAL*8, INTENT(out)   :: f
   REAL*8, INTENT(out), DIMENSION(1:n) :: g
   CALL VLS2fg(x,f,g)
END SUBROUTINE calfun

FUNCTION funval(n, x)
   IMPLICIT NONE
   INTEGER, INTENT(in) :: n
   REAL*8, INTENT(in), DIMENSION(1:n) :: x
   REAL*8 :: funval, f
   REAL*8, DIMENSION(1:2) :: g
   CALL VLS2fg(x,f,g); funval = f
END FUNCTION funval
```

An inequality constrained problem

Finally we give an example of a problem which has inequality constraints only. This is VLS2b and it uses OPTIMIZE4 as the interface to the OPTIMA procedures. Unlike OPTIMIZE3 this permits the use of the feasible-point methods B-SUMT and IPM. Any changes to be made to this example code should be done in the light of comments made in connection with the mixed equality-inequality constrained problem above.

```
PROGRAM VLS2b
! *** Main program for problem VLS2b ***
   USE minpac
   IMPLICIT NONE
   REAL*8, DIMENSION(:), ALLOCATABLE :: x
   CHARACTER*1 :: ans
   INTEGER :: n, method, m, me, i
   REAL*8:: t(5), z(5)
   COMMON/sumt/method
   INTERFACE
```

```
      SUBROUTINE calfun(x,n,m,f,c,g,A)
      USE opfad;
      IMPLICIT NONE
      INTEGER, INTENT(in) :: n,m
      REAL*8, INTENT(in), DIMENSION(1:n) :: x
      REAL*8, INTENT(out)  :: f
      REAL*8, INTENT(out), DIMENSION(1:m) :: c
      REAL*8, INTENT(out), DIMENSION(1:n), OPTIONAL :: g
      REAL*8, INTENT(out), DIMENSION(1:m,1:n), OPTIONAL :: A
      END SUBROUTINE calfun
      SUBROUTINE Sumtfun(n,x,P,gradP)
      USE minpac
      INTEGER, INTENT(in) :: n
      REAL*8, DIMENSION(1:n), INTENT(in) :: x
      REAL*8, INTENT(out) :: P
      REAL*8, DIMENSION(1:n), INTENT(out) :: gradP
      END SUBROUTINE Sumtfun
      END INTERFACE
      PRINT"(' Problem VLS2b with inequalities only')"
! *** number of variables, total number of constraints, number of equality
           constraints ***
      n = 2; m=10; me = 0
! *** Set starting guess ***
      ALLOCATE(x(1:n)); x= (/0.0, 0.0/);
      PRINT"(' Feasible starting point (y/n)?')"; READ*, ans
      IF(ans == 'y')x = (/0.925, -0.4712/)
      CALL print_minpac_history;  CALL set_minpac_iterations(5000);
      CALL set_initial_penalty(0.1D0); CALL set_penalty_scaling(0.25D0)
      CALL OPTIMIZE4(n,x,m,me,method,Sumtfun,calfun); DEALLOCATE(x)
      END PROGRAM VLS2b

      SUBROUTINE VLS2cfg(x,f,c,g,A)
! *** user-supplied function and constraints for problem VLS2b ***
      USE opfad
      REAL*8, DIMENSION(1:2), INTENT(in) :: x
      REAL*8, INTENT(out) :: f
      REAL*8, DIMENSION(1:10), INTENT(out) :: c
      REAL*8, DIMENSION(1:2), INTENT(out) :: g
      REAL*8, DIMENSION(1:10,1:2), INTENT(out) :: A
      TYPE(doublet) :: xx(2), ff, ss(5), cc(10)
      REAL*8 :: t(5),z(5)
      INTEGER :: k
      t = (/0.0,1.0,2.0,3.0,4.0/); z = (/1.0,0.5,0.4,0.3,0.2/) ! *** data points ***
      CALL INITIALIZE(2,x,xx) ! *** convert real variables to doublet form ***
      ff = 0.0D0
!  *** evaluate residuals and sum-of-squares function ***
      DO k = 1,5
         ss(k) = (xx(1)*exp(xx(2)*t(k)) - z(k));  ff = ff + ss(k)**2
      END DO
! *** calculate inequality constraints (upper and lower bounds on residuals)
      DO k = 1,5
      cc(k) = ss(k) + 0.08D0;  cc(k+5) = 0.08D0 - ss(k)
      END DO
      f = VALUE(ff);  g = GRADIENT(ff);  ! *** extract function and gradient ***
      c = VALUES(cc,10); A = NORMALS(cc,10) ! *** extract constraints and Jacobian
      END SUBROUTINE VLS2cfg
```

```
SUBROUTINE calfun(x,n,m,f,c,g,A)
  USE opfad;
  IMPLICIT NONE
  INTEGER, INTENT(in) :: n,m
  REAL*8, INTENT(in), DIMENSION(1:n) :: x
  REAL*8, INTENT(out)  :: f
  REAL*8, INTENT(out), DIMENSION(1:m) :: c
  REAL*8, INTENT(out), DIMENSION(1:n), OPTIONAL :: g
  REAL*8, INTENT(out), DIMENSION(1:m,1:n), OPTIONAL :: A
  REAL*8, DIMENSION(1:n) :: gdum
  REAL*8, DIMENSION(1:m,1:n) :: Adum
  CALL VLS2cfg(x,f,c,gdum,Adum)
  IF(PRESENT(g))g=gdum; IF(PRESENT(A))A=Adum
END SUBROUTINE calfun

SUBROUTINE Sumtfun(n,x,P,gradP)
 USE minpac
 INTEGER, INTENT(in) :: n
 REAL*8, DIMENSION(1:n), INTENT(in) :: x
 REAL*8, INTENT(out) :: P
 REAL*8, DIMENSION(1:n), INTENT(out) :: gradP
 REAL*8 :: f
 REAL*8, DIMENSION(1:n) :: g
! *** dimension of c and first dimension of A must be total number of
          constraints ***
 REAL*8, DIMENSION(1:10) :: c
 REAL*8, DIMENSION(1:10,1:n) :: A
 INTEGER :: method
 COMMON/sumt/method
 CALL VLS2cfg(x,f,c,g,A)
! *** parameter 5 of Make_P and Make_AL must be number of equality
          constraints ***
 IF (method == 1)CALL Make_P(f,g,c,A,0,P,gradP)
 IF (method == 2)CALL Make_AL(f,g,c,A,0,P,gradP)
 IF (method == 4)CALL Make_B(f,g,c,A,P,gradP)
END SUBROUTINE Sumtfun
FUNCTION funval(n,y)
funval = 0.0
END FUNCTION funval
```

25.4. Modifying optimization methods

Some of the exercises in the main text invite the reader to make changes
to one of the OPTIMA algorithms. An example would be replacement
of the BFGS updating formula with the DFP one. Such tasks are not
particularly difficult but probably require a user to be a fairly confident
Fortran programmer. We give a few illustrations to help the reader
make some of the possible changes to algorithms that are suggested in
the text. With these illustrations as a guide, an enthusiastic reader with
a knowledge of Fortran should be able to identify other possibilities for
modifying – and possibly improving – the OPTIMA implementations.

Changing the update in a quasi-Newton method

This particular change can be made by adding a subroutine DFP to the module MINPAC which is similar to the existing BFGS routine listed below. The other change needed is the replacement of the CALL BFGS statement in the subroutine quasi-Newton with the corresponding CALL DFP statement.

```
SUBROUTINE BFGS(n,H,p,g,gold)
! BFGS update for inverse Hessian
   INTEGER, INTENT (in) :: n
   REAL*8, INTENT(inout), DIMENSION(1:n,1:n) :: H
   REAL*8, INTENT(in), DIMENSION(1:n) :: p,g,gold
   REAL*8, DIMENSION(1:n) :: y, Hy
   REAL*8 :: dy, yHy, temp, dnm, ynm
   INTEGER :: i,j
   y = g - gold;  dy = DOT_PRODUCT(p,y)
   dnm = SQRT(DOT_PRODUCT(p,p)); ynm = SQRT(DOT_PRODUCT(y,y))
   IF (dy <= 0.01*dnm*ynm) RETURN
   Hy = MATMUL(H,y);  yHy = DOT_PRODUCT(y,Hy); temp = (1.0D0 + yHy/dy)/dy
   DO i = 1,n
      DO j = i,n
         H(i,j) = H(i,j) - (Hy(i)*p(j) + Hy(j)*p(i))/dy + temp*p(i)*p(j)
         H(j,i) = H(i,j)
      END DO
   END DO
END SUBROUTINE BFGS
```

Changing the conjugate gradient search direction

We can produce a version of the conjugate gradient method which uses the Polak-Ribiere form of recurrence relation to obtain the search directions. In the OPTIMA subroutine Conjugate_Gradient we can replace the statement beta = dot_product(g,g)/dot_product(gold,gold) by one which calculates β using (11.2.1).

Changing the form of the barrier function

If we wish to change the B-SUMT method to use the reciprocal, rather than the logarithmic, barrier term then, in the subroutine Make_B, it is sufficient simply to replace the statement B = B - rbar*log(c(i)) by B = B + rbar/c(i).

References

1. L. Armijo, Minimization of Functions Having Continuous Partial Derivatives, Pacific J. Maths, **16**, pp 1–3, 1966.
2. M. Bartholomew-Biggs, *Anglicised by Common Use*, Waldean Press, 1998.
3. M. Bartholomew-Biggs, *Inklings of Complicity*, Pikestaff Press, 2003.
4. M. Bartholomew-Biggs, *Other Poetry*, II/24, 2003.
5. M. Bartholomew-Biggs, *The SHOp*, 20, 2006.
6. M.C. Bartholomew-Biggs, A Newton Method with a Two-dimensional Line Search, Advanced Modeling and Optimization, www.ici.ro/camo/journal **5**, pp 223–245, 2003.
7. M.C. Bartholomew-Biggs, IP from an SQP point of view, Optimization Methods and Software, **16**, pp 69–84, 2001.
8. M.C. Bartholomew-Biggs, Recursive Quadratic Programming Methods Based on the Augmented Lagrangian, Math. Prog. Study **31**, pp 21–41, 1987.
9. M.C. Bartholomew-Biggs, S.C. Parkhurst & S.P. Wilson, Global Optimization Approaches to an Aircraft Routing Problem, European Journal of Operational Research, **146**, pp 417–431, 2003.
10. M.C. Biggs, Constrained Minimization using Recursive Equality Quadratic Programming, in: F.A. Lootsma, (Ed), *Numerical Methods in Nonlinear Optimization*, Academic Press, 1972.
11. C.G. Broyden and M.T. Vespucci, *Krylov Solvers for Linear Algebraic Systems*, Studies in Computational Mathematics **11**, Elsevier, 2004.
12. C.G. Broyden, The Convergence of a Class of Double Rank Minimization Algorithms, Part 1, J. Inst. Maths. Appl. **6**, pp 76–90, 1970 and Part 2, J. Inst. Maths. Appl. **6**, pp 222–231, 1970.
13. R.H. Byrd, M. Hribar and J. Nodedal, An Interior Point Algorithm for Large Scale Nonlinear programming, OTC Technical Report 97/05, Optimization Technology Center, 1997.
14. A.R. Conn, N.I.M. Gould and Ph. L. Toint *Trust Region Methods*, MPS-SIAM Series on Optimization, Philadelphia, 2000.
15. W.C. Davidon, Variable-metric Method for Minimization, AEC Report ANL5990, Argonne National Laboratory, 1959.
16. R. Dembo, S. Eisenstat and T. Steihaug, Inexact Newton Methods, SIAM J. Numerical Analysis **10**, pp 400–408, 1982.

17. J.E. Dennis and R.B. Schnabel, *Numerical Methods for Unconstrained Optimization and Nonlinear Equations*, Prentice-Hall, 1983.
18. L.C.W. Dixon, Quasi-Newton Algorithms Generate Identical Points, Math. Prog. **2**, pp 383–387, 1972.
19. L.C.W. Dixon, Quasi-Newton Algorithms Generate Identical Points, Part 2 – Proofs of Four New Theorems, Math. Prog. **3**, pp 345–358, 1972.
20. A.V. Fiacco and G.P. McCormick, *Nonlinear Programming – Sequential Unconstrained Minimization Techniques*, John Wiley, 1968. Reissued by SIAM Classics in Applied Mathematics, 1990.
21. R. Fletcher and C. Reeves, Function Minimization by Conjugate Gradients, Comp. J. **7**, pp 149–154, 1964.
22. R. Fletcher and M.J.D. Powell, A Rapidly Convergent Descent Method for Minimization, Computer J. **6**, pp 163–168, 1963
23. R. Fletcher and S. Leyffer, Nonlinear Programming without a Penalty Function, Math. Prog. **91**, pp 239–269, 2002.
24. R. Fletcher and S.A. Lill, A Class of Methods for Nonlinear programming II: Computational Experience. in: J.B. Rosen, O.L. Mangasarian and K. Ritter, (Eds), *Nonlinear Programming*, Academic Press, 1972.
25. R. Fletcher, A General Quadratic Programming Algorithm, J. Inst. Maths Appl. **7**, pp 76–91, 1971
26. C.A. Floudas and P.M. Pardolos, (Eds), *Encyclopedia of Optimization*, Kluwer Academic, 2001.
27. A.B. Forbes, M. Bartholomew-Biggs and B.P. Butler, Optimization algorithms for generalized distance regression in metrology, in: P. Ciarlini, A.B. Forbes, F. Pavese, D. Richter (Eds.), *Advanced Mathematical and Computational Tools in Metrology IV*, Series on Advances in Mathematics for Applied Sciences, Vol. 53, World Scientific, Singapore, 2000, pp 21–31.
28. A. Forsgren and P.E. Gill, Primal-dual Interior Point Methods for Nonconvex Nonlinear Programming, SIAM J. Opt. **8**, pp 1132–1152, 1998.
29. Frontline Systems Inc., www.solver.com
30. P.E. Gill and W. Murray, Newton-type Methods for Unconstrained and Linearly Constrained Optimization, Mathematical Programming **30**, pp 176–195, 1974.
31. N.I.M. Gould and Ph. L. Toint, SQP Methods for Large-scale Nonlinear Programming, in: M.J.D. Powell and S. Scholtes, (Eds), *System Modelling and Optimization: Methods, Theory and Applications*, Kluwer, 1999.
32. A Griewank, *Evaluating Derivatives: Principles and Techniques of Automatic Differentiation*, SIAM, 2000.
33. S.P. Han, Superlinearly Convergent Variable-metric Algorithms for General Nonlinear Programming Problems, Math. Prog., **11**, pp 263–282, 1976.
34. S.E. Hersom, The Practice of Optimization, in: L.C.W. Dixon and G.P.Szego, (Eds), *Towards Global Optimization*, North-Holland, 1975
35. M.R. Hestenes and E.L. Stiefel, Methods of Conjugate Gradients for Solving Linear Systems, J. Res. Nat. Bureau of Standards **49**, pp 409–436, 1952.
36. N.J. Higham, *Accuracy and Stability of Numerical Algorithms*, SIAM, Philadelphia, 1996.
37. R. Hooke and T.A. Jeeves, Direct Search Solution of Numerical and Statistical Problems, J.ACM **8**, pp 212–229, 1961.
38. A. Jennings and J.J. McKeown, *Matrix Computation*, Second Edition, John Wiley, 1992.

39. D.R.Jones, C.D. Perttunen and B.E. Stuckman, Lipschitzian Optimization without the Lipschitz Constant, J. Opt. Theory & Appl, **79**, pp 157–181, 1993

40. N. Karmarkar, A New Polynomial Time Algorithm for Linear Programming, Combinatorica **4**, pp 373–395, 1984.

41. M. Kijima, H. Marimura and Y Suzuki, Periodical Replacement Problem without Assuming Minimal Repair, European J. Operational Research, **37**(2), 194–203, 1988.

42. M. Kijima, Some Results for Repairable Systems with General Repair, J. Applied Probability **26**, 89–102, 1989.

43. L.S. Lasdon, A.D. Waren, A. Jain and M. Ratner, Design and Testing of a Generalised Reduced Gradient Code for Nonlinear Programming, ACM Trans. Math. Soft. **4**, pp 34–50, 1978.

44. K. Levenberg, A Method for the Solution of Certain Nonlinear Problems in Least Squares, Quart. Appl. Maths., **2**, pp 164–168, 1944.

45. J. Lucas, (Ed), *Take Five*, Shoestring Press, 2003.

46. N. Maratos, Exact Penalty Function Algorithms for Finite-dimensional and Control Optimization Problems, PhD Thesis, London University, 1978

47. D.W. Marquardt, An Algorithm for Least Squares Estimation of Nonlinear Parameters, SIAM J., **11**, pp 111–115, 1963.

48. Microsoft Corporation, www.microsoft.com

49. W. Murray An Algorithm for Constrained Optimization, in: R. Fletcher, (Ed), *Optimization*, Academic Press, 1969.

50. J.A. Nelder and R. Mead, A Simplex Method for Function Minimization, Comp. J. **7**, pp 308–313, 1965.

51. E. Polak and G. Ribiere, Note sur la Convergence de Methode de Directions Conjugées, Revue France Inform. Rech. Oper. **16**, pp 35–43, 1969.

52. M.J.D. Powell, A Fast Algorithm for Nonlinearly Constrained Optimization Calculations, in: G. Watson (Ed), *Numerical Analysis, Dundee 1977*, Vol 630 of Lecture Notes in Mathematics, Springer, 1978.

53. M.J.D. Powell, A Method for Nonlinear Constraints in Minimization Problems, in: R. Fletcher (Ed), *Optimization*, Academic Press, 1969.

54. M.J.D. Powell, On the Convergence of the Variable-metric Algorithm, J. Inst. Maths. Appl. **7**, pp 21–36, 1971.

55. M.J.D. Powell, Some Global Convergence Properties of a Variable-metric Algorithm without Line Searches, in: R.W. Cottle and C.E. Lemke (Eds.), *Nonlinear Programming* AMS, 1976.

56. D. Pu, The Convergence of Broyden Algorithms without Convexity Assumption, System Science & Maths Science **10**, pp 289–298, 1997.

57. A. Rinnooy-Kan and G. Timmer, Stochastic Global Optimization Methods Part I : Clustering Methods, Math. Prog. **39**, pp 27–56, 1987.

58. A. Rinnooy-Kan and G. Timmer, Stochastic Global Optimization Methods Part II : Multi-level Methods, Math. Prog. **39**, pp 57–78, 1987.

59. R.T. Rockafellar, A Dual Approach to Solving Nonlinear Programming Problems using Unconstrained Optimization, Math. Prog, **5**, pp 354–373, 1973.

60. R. Schnabel and E. Eskow, A New Modified Cholesky Factorization, SIAM J. Scientific Computing **11**, pp 1136-1158, 1991.

61. The Mathworks Inc., www.mathworks.com

62. The Numerical Algorithms Group, http://www.nag.co.uk

63. R.J. Vanderbei *Linear Programming: Foundations and Extensions*, Kluwer Academic, 1996.

64. R.B. Wilson, A Simplicial Method for Concave Programming, PhD Dissertation, Harvard University, Cambridge MA, 1963.

65. S.P. Wilson, Aircraft Routing using Nonlinear Global Optimization, PhD Thesis, University of Hertfordshire, 2003.

66. P. Wolfe, Convergence Conditions for Ascent Methods, SIAM Review, **11**, pp 226–235, 1969.

67. S.J. Wright, Recent Developments in Interior Point Methods, in: M.J.D. Powell and S. Scholtes, (Eds), *System Modelling and Optimization: Methods, Theory and Applications*, Kluwer, 1999.

68. J.Z. Zhang and C.X. Xu, A Class of Indefinite Dog-leg Methods for Unconstrained Minimization, SIAM J. Optim. **9**, pp 646–667, 1999.

Index